Further Principles and Systems for Radio & TV Mechanics

D1799875

Further Principles and Systems for Radio & TV Mechanics

FULL COVERAGE OF C & G 222 PART 2

K F IBRAHIM

BSc (Electrical Engineering)
Lecturer, Willesden College of Technology

PITMAN PUBLISHING

PITMAN PUBLISHING LIMITED
39 Parker Street, London WC2B 5PB

Associated Companies

Copp Clark Ltd, Toronto
Fearon Publishers Inc, Belmont, California
Pitman Publishing New Zealand Ltd, Wellington
Pitman Publishing Pty Ltd, Melbourne

© K F Ibrahim 1978

First published in Great Britain 1978

All rights reserved. No part of this publication
may be reproduced, stored in a retrieval system,
or transmitted, in any form or by any means,
electronic, mechanical, photocopying, recording
and/or otherwise without the prior permission of
the publishers. This book may not be lent, resold,
hired out or otherwise disposed of by way of trade
in any form of binding or cover other than that in
which it is published, without the prior consent
of the publishers. This book is sold subject to
the Standard Conditions of Sale of Net Books and
may not be resold in the UK below the net price.

Camera copy prepared by Morgan-Westley **M**

Printed by photolithography and bound in Great
Britain at The Pitman Press, Bath

ISBN 0 273 01122 7

PREFACE

This book is a follow-up to *Principles and Systems for Radio and TV Mechanics* and covers the whole syllabus for part II of the City and Guilds 222 course.

It is divided into three parts: Principles, Systems, and Practical Techniques. Apart from sections 13 and 14, which deal with thermionic theory and devices, the part on Systems covers the basic principles of semiconductor devices and their various applications in modern electronics. Although the treatment is non-mathematical, the circuits are explained in depth to give the students a sound understanding in electronics. Wherever applicable, I have indicated typical values of components that may be found in modern electronic equipments.

The part on Practical Techniques deals with some important practical aspects of electronics, ending with a section on fault diagnosis. Here I have departed from the academic approach to fault finding, so far common in textbooks, in which the components in a circuit such as an amplifier stage are considered individually, noting the effect of each one when it goes open circuit, short circuit, etc., and resulting in a long list of faults for each circuit. Instead I have employed a dynamic approach where the state (e.g. cut-off or saturation) of the circuit under test is examined as a first step to identifying the faulty component.

It has been my intention throughout the book to meet the challenge of conveying complex electronic concepts and systems in a manner that may be understood by the student or the engineer who has not had the opportunity of a mathematical background, but who may nonetheless wish to acquire the necessary skills to service modern electronic equipment. To this end I hope I have succeeded. In this connection too I wish to

thank the students with whom I have been associated whose inquisitive minds contributed to the
writing of this book.

Apart from the students taking the City and
Guilds 222 part II examination, this book should
be very useful for TEC students at levels II and
III taking electronic engineering. It should
also be useful to practising engineers who would
like to back their practical background with
theoretical knowledge.

Finally I wish to thank June Kingston who,
over a period of nine months tirelessly typed
the script in her spare time.

K.F.I.

CONTENTS

SYSTEMS

(xii)

PRACTICAL TECHNIQUES

1 Mechanics and Light

Revision

The application of a force over a distance constitutes a measure of *work*.

Work = Force × Distance   or $W = F \times d$

The unit of force is the Newton N, the unit for distance is the metre m, and the unit for work is the Joule J.

Power is defined as the rate of doing work. Hence

$$\text{Power} = \frac{\text{Work}}{\text{Time}} \quad \text{in watts W}$$

In other words, 1 W = 1 Joule/second.

The kilowatt-hour kWh is a convenient unit of work used in the power generation industry as a tariff unit. 1 kWh is the work done by a machine having a power of 1 kW over a period of one hour.

1 kW = 1000 W = 1000 J/s

Hence, over a period of one hour the work done by a 1 kW machine is

1 kWh = 1000 × 60 × 60 = 3 600 000 = 3.6×10^6 J

Stress and Strain

When a force is applied to a piece of material such as a wire or a rod, the material is said to be under stress. Stress is defined as

$$\text{Stress} = \frac{\text{Applied force or load}}{\text{Cross-sectional area}} \; \text{N/cm}^2$$

For example, a metal bar having a cross-sectional area a of 2 cm^2 holding a weight of 100 N has a stress of

$$\frac{\text{Load}}{a} = \frac{100 \text{ N}}{2 \text{ cm}^2} = 50 \text{ N/cm}^2$$

The load or force is sometimes quoted in kilogramme-force kgf, i.e. the force of one kilogramme.

1 kgf = 9.81 N

When a piece of metal such as a rod is loaded by a force, it begins to expand, i.e. increase its length. Strain is defined as the ratio

$$\text{Strain} = \frac{\text{Change in length}}{\text{Original length}} \quad \text{(no units)}.$$

For example, if a 2 m long wire increases its length by
1 μm (1 micron) = 10^{-6} m when under stress, then the
strain experienced by the rod is

$$\frac{1 \ \mu m}{2 \ m} = \frac{10^{-6} \ m}{2 \ m} = 0.5 \times 10^{-6} \ m = 0.5 \ \mu m$$

The change in length may be a reduction of the original
length where a piece of metal is subjected to *compressive*
stress as opposed to the *tensile* stress considered above.

Velocity

An object that travels a distance d in time t is said to
have a higher speed than a second object that travels the
same distance in longer time. Speed or velocity is
defined as

$$\text{Velocity} = \frac{\text{Distance covered}}{\text{Time taken}} = \frac{d}{t}$$

where d is in metres, t in seconds, and velocity in m/s.
For example, a car that takes 1 min to travel a distance
of 100 m has a speed of

$$\frac{100 \ m}{1 \ min} = \frac{100 \ m}{60 \ s} = 1.67 \ m/s$$

Acceleration

A car that changes its speed from v_1 to v_2 in time t is
said to have a greater acceleration than a second car
which requires longer time to achieve the same change in
speed. Acceleration is defined as the rate of change of
velocity, that is

$$\frac{\text{Change in velocity}}{\text{Time taken}} = \frac{v_2 - v_1}{t} \ m/s^2$$

For example if an object travelling at 2 m/s reaches a
speed of 12 m/s after 5 s, it then has an acceleration of

$$\frac{12 - 2}{5} = \frac{10}{5} = 2 \ m/s^2$$

Rotary Movement

When a rod OA rotates about a centre O as shown in Fig.
1.1, it has what is known as angular velocity. *Angular
velocity* is a measure of the speed of rotation. It is
defined as

$$\text{Angular velocity} = \frac{\text{The angle covered}}{\text{Time taken}}$$

For instance, if OA rotates to OB covering an angle θ in
time t, then the angular velocity ω is

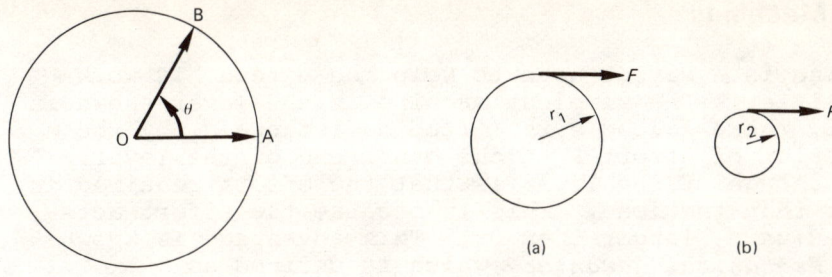

Fig.1.1   Fig.1.2 Torque = force × radius

$$\omega = \frac{\theta}{t}$$

The unit for angle is the radian where one complete revolution $360^{\circ} = 2\pi$ radians. Hence the unit for angular velocity is radians/second or rad/s. For example, if a wheel completes 2 revolutions every second then the angle covered by the wheel is $2 \times 2\pi = 4\pi$ rad, giving an angular velocity $\omega = 4/1 = 4$ rad/s.

Example What is the angular velocity of a lever rotating at a speed of 10 rpm (revolutions per minute)?

Solution Angle covered $= 10 \times 2\pi = 20\pi$ rad. Time taken $= 1$ min $= 60$ s.

Hence, angular velocity $\omega = \frac{20\pi}{60} = \frac{\pi}{3}$ rad/s or 1.04 rad/s.

The Flywheel

Flywheels are used to smooth out variations in the speed of rotation thus keeping the angular velocity constant. They are used for instance in tape decks and radiogram turntables. Flywheels are constructed with a heavy rim and a light centre in order to ensure high kinetic energy, making the flywheel more effective.

Torque

If a force F is applied to the wheel shown in Fig.1.2a at a radius r_1, the force will try to rotate the wheel in the clockwise direction. The rotating force is dependent on both the force F itself as well as the radius r_1. For instance the same force F has a smaller turning effect if applied at a shorter radius r_2 as shown in 1.2b. This turning force is known as the torque.

Torque = Force × Radius = $F \times r$

With force in newtons and the radius in metres, torque is given the unit newton-metre, Nm.

Simple Machines

A machine is a device used to help man extend his work capabilities. The simplest machine is the lever shown in Fig.1.3, where a load M is lifted a distance d_1 by the application of effort F at the other end of the lever. The advantage of the lever is that the effort required is smaller than the load. This is because the effort acts at a radius r_2 larger than r_1. This advantage is known as the *Mechanical Advantage* which is defined as

$$\text{Mechanical advantage} = \frac{\text{Load}}{\text{Effort}}$$

As can be seen, the distance moved by the effort is greater than the distance moved by the load. Hence the effort and the load move at different velocities giving a

$$\text{Velocity Ratio} = \frac{\text{Velocity of effort}}{\text{Velocity of load}}$$

Since the time taken by the effort and the load to cover their respective distances is the same then

$$\text{Velocity ratio} = \frac{\text{Distance moved by the effort}}{\text{Distance moved by the load}} = \frac{d_1}{d_2}$$

Pulleys

Fig.1.4 shows one pulley used to lift a load M. It can be seen from 1.4b that, if the effort F pulls the rope by 2 cm, the load is moved by half that distance, namely 1 cm. Thus the effort required is approximately half the load. Multiple pulleys may be used to improve the mechanical advantage.

Gears

Fig.1.5 shows two meshed gear wheels. Gear wheel A is the *driver* which provides the effort. Gear wheel B is the *driven* wheel which provides the load.

$$\text{Velocity ratio} = \frac{\text{Number of revolutions of driver wheel}}{\text{Number of revolutions of driven wheel}}$$

$$= \frac{\text{Number of teeth of driven wheel}}{\text{Number of teeth of driver wheel}}$$

$$= \frac{n_2}{n_1}$$

For example, if the number of teeth of the driver and driven wheels are 15 and 30 respectively then the velocity ratio is 30/15 = 2.

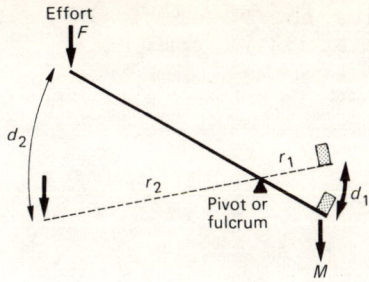

Fig.1.3 The simple lever

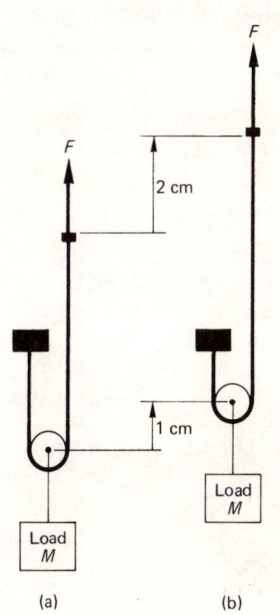

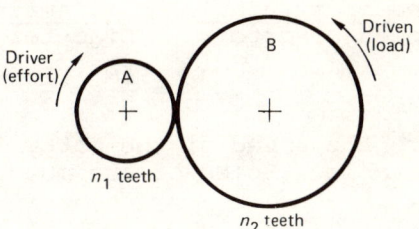

Fig.1.5 Two meshed gear wheels

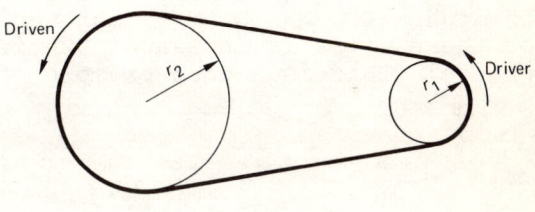

(a)   (b)

Fig.1.4 The simple pulley   *Fig.1.6*

A similar effect may be produced by using a belt to transfer the effort from the driver to the driven wheel as shown in Fig.1.6. In this case however, the rotation of the driver and the driven is in the same direction. The velocity ratio may now be calculated from the diameter or the radius ratios.

$$\text{Velocity ratio} = \frac{\text{Radius of driven wheel}}{\text{Radius of driver wheel}} = \frac{r_2}{r_1}$$

Example Refer to Fig.1.7. Calculate the speed of the motor if the turntable is to rotate at 33 $\frac{1}{3}$ rpm.

Solution

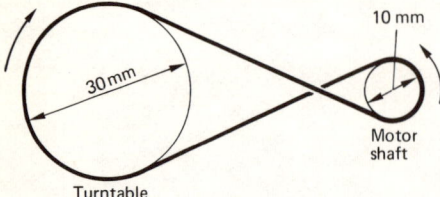

Fig.1.7

$$\frac{\text{Speed of turntable}}{\text{Speed of motor}} = \frac{\text{Diameter of motor shaft}}{\text{Diameter of turntable shaft}}$$

$$= \frac{10}{30} = \frac{1}{3}$$

Therefore the speed of the motor is three times that of the turntable. Speed of the motor = 3 × 33 $\frac{1}{3}$ = 100 rpm.

Light and Colour

Electromagnetic waves extend from low frequencies (long waves) through the radio frequencies, infra-red, light waves, ultra violet to X-rays and cosmic rays (Fig.1.8). All electromagnetic waves consist of an electric field at right angles to a magnetic field travelling at the speed of light (300 000 000 m/s) through space. As the frequency of these waves is increased, their properties begin to change. This is a good example of a quantitative change

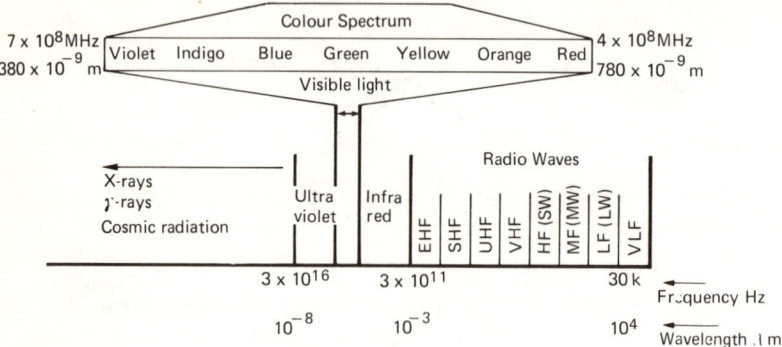

Fig.1.8 *Electromagnetic wave spectrum*

(an increase in frequency) leading to a qualitative change (change in property). For example, the method of propagation of low-frequency electromagnetic waves is different from that at higher frequencies, e.g. v.h.f. and u.h.f. (hundreds × MHz) have entirely different qualities from those of short waves on the one hand and X-rays on the

other. In a small band within the electromagnetic spectrum (between 4×10^8 and 7×10^8 MHz) electromagnetic
waves become visible light.

Every point on the visible spectrum corresponds to a
certain wavelength (the distance in space between two
adjacent peaks of the wave, also see p.175), and every one
of these wavelengths (or frequencies) is perceived by the
eye as a certain tint (or hue), with red at the low-
frequency (long wavelength) end through to violet at the
high-frequency (shorter wavelength) at the other end.

Pure or Spectral Colours

Spectral or pure colours are those colours that appear on
the colour spectrum (Fig.1.8) or rainbow. It is possible
to produce spectral colours (e.g. yellow) by combining
two other spectral colours, one having a shorter wavelength, i.e. higher frequency (green) and another having
a longer wavelength, i.e. lower frequency (red).

The colour of an object is determined by certain chem-
icals known as pigments which absorb some spectral colours
while reflecting others. For example, when white colour
falls on a red object, all the spectral colours with the
exception of red are absorbed by the pigments. Red is
reflected to create the sensation of colour in the eye.

Saturation and Hue

The colour of any object contains two qualities: (1) saturation which indicates the proportion of pure colour
content and (2) hue which indicates the primary component
of the colour of the object. For example, a desaturated
red is pure red mixed with white to make pink. Pink,
therefore, has a red hue.

Light Units and Definitions

Consider a light source B (e.g. an electric bulb) emitting
light upon a surface A from a distance d (Fig.1.9). The
strength of the light may be measured by the power of the

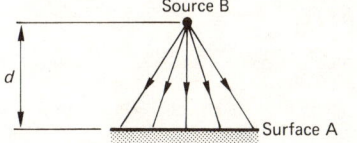

Fig.1.9

light bulb (luminous intensity) or by the strength of the
light waves (luminous flux) or by the concentration of
light upon the surface (illumination). More precisely,
luminous intensity measured in candela (cd) describes the
power of the light source; *luminous flux* in lumen (lm)
measures the flow of light per second; and *illumination*

in lux (ℓ) measures the concentration of luminous flux upon a surface, where 1 lux = 1 lumen/square metre.

Luminance is another term used specially in TV systems to give a measure of the light emitted <u>from</u> a surface. If the light source in Fig.1.9 is moved away from the surface then the illumination of the surface will decrease, and vice versa. The illumination is proportional to the square of the distance between the source and the surface. This is known as the *inverse square law* of illumination.

For example, a source of 1 candela at a distance of 1 m produces an illumination of the surface of 1 lux. At a distance of 2 m the illumination is reduced by a factor of $1/2^2 = 1/4$. On the other hand, if the distance is reduced to $\frac{1}{2}$ m = 0.5 m, then the illumination is

$$\frac{1}{(0.5)^2} = \frac{1}{0.25} = 4 \text{ lux}$$

2 D.C. Networks

Revision

Power dissipated in a resistor is voltage V × current I or $I^2 \times R$ or V^2/R in watts.

RESISTORS IN SERIES (Fig.2.1) Total resistance = $R_1 + R_2$.

Total voltage $V_T = V_1 + V_2$

$$V_1 = \frac{V_T}{R_1 + R_2} \times R_1 \quad \text{and} \quad V_2 = \frac{V_T}{R_1 + R_2} \times R_2$$

For two equal resistors in series, the voltage across each resistor is half the total voltage.

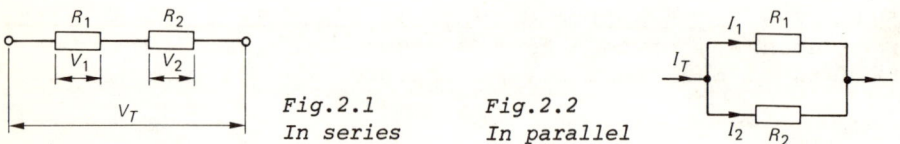

Fig.2.1
In series

Fig.2.2
In parallel

RESISTORS IN PARALLEL (Fig.2.2)

$$\text{Total resistance} = \frac{R_1 \times R_2}{R_1 + R_2}$$

The total resistance of two equal resistors in parallel is half one of them.

Total voltage = voltage across R_1 = voltage across R_2.

Total current $I_T = I_1 + I_2$.

Example (1)   Calculate the voltage across R_3 in Fig.2.3.
Solution   Total resistance between A and B = ½ × 12 = 6Ω.
   Voltage across R_3 = voltage between A and B = V_{AB} = 4.5 V.

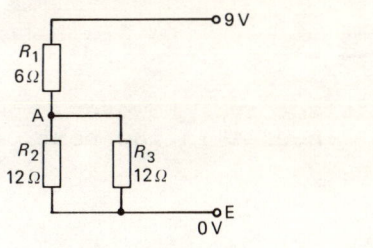

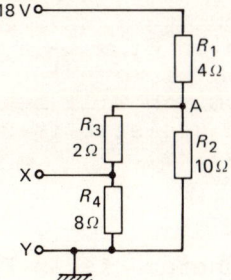

Fig.2.3

Fig.2.4

Example (2)   Calculate the voltage between X and Y in Fig.2.4.
Solution $R_3 + R_4 = 2 + 8 = 10Ω$.

   Total resistance between A and Y = 10/2 = 5 Ω.

Hence, voltage at point A, $V_{AY} = \dfrac{18}{4 + 5} \times 5 = \dfrac{18}{9} \times 5$

$$= 10 \text{ V}$$

Voltage across X and Y, $V_{XY} = \dfrac{10}{2 + 8} \times 8 = \dfrac{10}{10} \times 8 = 8 \text{ V}$

Example (3)   A 2.5Ω resistor is connected across a 50 V supply.

The power dissipated in the resistor = $V^2/R = \dfrac{50 \times 50}{2.5}$

$$= 1000 \text{ W} = 1 \text{ kW}$$

Example (4)   Refer to Fig.2.5. If the voltage reading at point A is 12 V when switch S is open, what is the reading if the switch is closed?
Solution   When S is closed the total resistance of the circuit is 0.1 + 4.9 = 5Ω.
   Thus the current through the circuit is 15/5 = 3 A, giving a voltage across the 4.9Ω load resistor of 3 × 4.9 = 14.7 V, which is the voltage at point A.

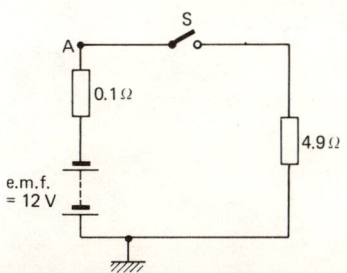

Fig.2.5

Example (5)   State the minimum power rating of resistor R_1 in Fig.2.6.
Solution   Total voltage across the circuit = 9 - 5 = 4 V.
   Voltage across R_1 = ½ × 4 = 2 V.

$$\text{Power dissipated in } R_1 = \frac{V^2}{R_1} = \frac{2 \times 2}{10} = 0.4 \text{ W.}$$

A power rating of 0.4 W will be satisfactory. However, 0.5 W or ½ W rated resistor is used since it is the next possible nominal value.

Capacitor Networks

Charge $Q = C \times V$ where C is in Farads and V is in Volts giving Q in coulombs.
   For two capacitors in *parallel* the total capacitance

$$C_T = C_1 + C_2$$

For two capacitors in *series* (Fig.2.7)

$$\text{Total capacitance} = \frac{C_1 \times C_2}{C_1 + C_2}$$

So that the total capacitance of two equal capacitors in series is half the value of one of them.

$$\text{Total voltage } V_T = V_1 + V_2 \text{ with the ratio } \frac{V_1}{V_2} = \frac{C_2}{C_1}$$

In general, when a number of capacitors are connected in series across a d.c. supply, the voltage across each capacitor is inversely proportional to its capacitance. For two capacitors in series (Fig.2.7) the voltage across C_1 is given as

$$V_1 = \frac{V_T}{C_1 + C_2} \times C_2$$

Example (1)   In Fig.2.8 the total capacitance is calculated as follows. Total capacitance of parallel combination is

$$C_2 + C_3 = 10 + 20 = 30 \text{ pF}$$

Since C_1 is also 30 pF then the total capacitance is ½ × 30 = 15 pF.

Example (2)   In Fig.2.9 the voltage across capacitor C_1 is

$$\frac{V_T}{C_1 + C_2} \times C_2 = \frac{30}{0.1 + 0.2} \times 0.2 = 20 \text{ V}$$

giving a voltage across C_2 = 30 - 20 = 10 V.

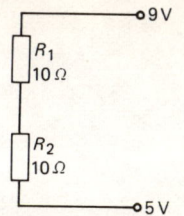

Fig.2.6

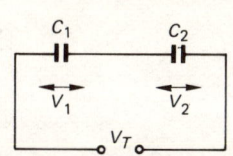

Fig.2.7 Capacitors in
series

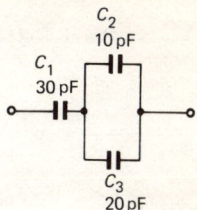

Fig.2.8

Working Voltage

Every capacitor has a maximum voltage it can sustain
before the dielectric breaks down. This voltage is known
as the working voltage of the capacitor and it must never
be exceeded. When used in a.c. applications, the peak
voltage applied across the capacitor must not exceed the
specified working voltage of the capacitor. The working
voltage of a number of capacitors connected in parallel
is that of the capacitor with the lowest working voltage.
For example in Fig.2.10, the working voltage of the
parallel circuit is 25 V.

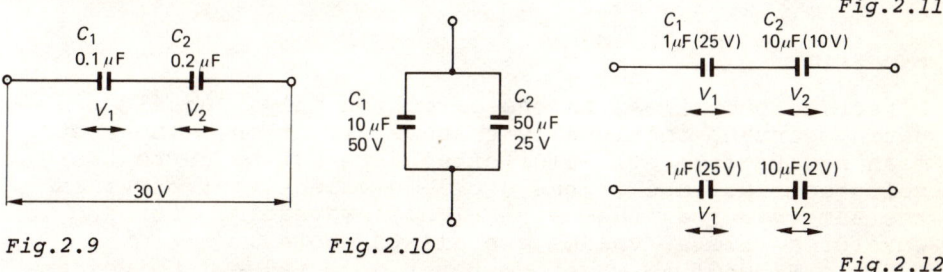

Fig.2.9

Fig.2.10

Fig.2.11

Fig.2.12

For capacitors in series the working voltage is more
difficult to work out. Consider the circuit in Fig.2.11
where C_1 (1 µF, working voltage WV = 25 V) is in series
with C_2 (10 µF, WV = 10 V). Since the smaller capacitor
C_1 will sustain a higher voltage than C_2, then start the
calculations with C_1 at its specified working voltage of
25 V. Thus $V_1 = 25$ V. But $V_2/V_1 = C_1/C_2$, so

$$V_2 = V_1 \times \frac{C_1}{C_2} = 25 \text{ V} \times \frac{1 \text{ µF}}{10 \text{ µF}} = 2.5 \text{ V}$$

Since V_2 is below the working voltage of C_2, then the
overall working voltage is 25 + 2.5 = 27.5 V.

However, if C_2 had a working voltage of say 2 V as in
Fig.2.12, it will reach its working voltage before C_1
voltage reaches 25 V. The overall working voltage is
calculated as follows:

$$V_2 = 2 \text{ V} \quad \text{and} \quad V_1 = V_2 \times \frac{C_2}{C_1} = 2 \times \frac{10}{1} = 20 \text{ V}$$

giving an overall working voltage of 20 + 2 = 22 V.

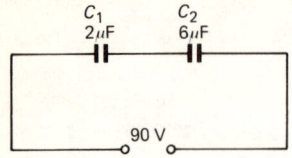

Fig.2.13

Example   Refer to Fig.2.13 with C_1 and C_2 having the same working voltage.   What is the minimum working voltage that may be used?
Solution   Since C_1 sustains a higher voltage across it than C_2, the former will reach its working voltage first.

$$V_1 = \frac{90}{C_1 + C_2} \times C_2 = \frac{90}{2 + 6} \times 6 = \frac{90}{8} \times 6 = 60 \text{ V}$$

The minimum working voltage is, therefore, 60 V.

3 A.C. Theory

Revision

Direct current flows in one direction only.   An alternating current continually changes direction.   The root mean square (r.m.s.) value of an alternating current or voltage is defined as the d.c. equivalent that provides the same *power* as the original a.c. waveform.   For a.c. waveforms, r.m.s. values are always quoted.
   For a SINUSOIDAL waveform the r.m.s. value is given as

   r.m.s. = 0.707 × peak value

R.M.S. for Complex Waveforms

The above formula is true only for a sinusoidal waveform. R.m.s. values for other waveforms may be calculated as follows:
(1) Square the waveform over one cycle.   Note that squaring a negative value turns it into a positive value.
(2) Take the mean of the squared waveform.
(3) Take the square root of the mean of the squared waveform, which gives the root mean square.
Hence, using the SQUARE wave shown in Fig.3.1a, squaring the positive half cycle gives $3 \times 3 = 9$.   Squaring the negative half cycle gives $(-3) \times (-3) = 9$.   The average over the squared cycle is therefore 9.   Hence the r.m.s. voltage is $\sqrt{9} = 3$ V.
   Comparing this with a sine wave having positive and negative peaks of +3 V and -3 V respectively, with an

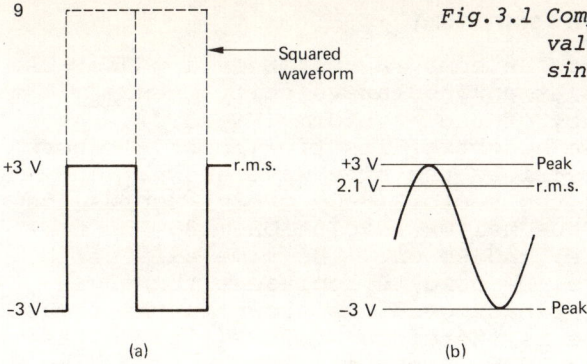

Fig.3.1 *Comparison between the r.m.s.*
values of a square wave and a
sine wave

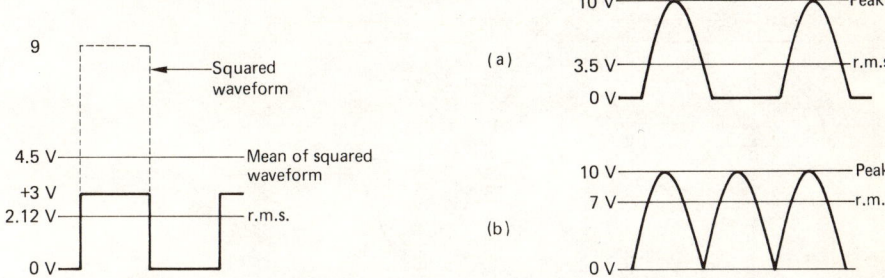

Fig.3.2 R.M.S. for positive
going square wave

Fig.3.3 (a) *The r.m.s. value of a half-wave*
rectified sine wave is ½ × 0.7 × peak
(b) *The r.m.s. value of a full-wave*
rectified sine wave is 0.7 × peak

r.m.s. of 0.707 × 3 = 2.1 V (Fig.3.1b), we find that the
square wave has a higher r.m.s. value than the sine wave.
This is because the area under the square waveform is
larger than that under the sinusoidal waveform when both
are alternating between the same positive and negative
values.

(Fig.3.2 shows a square wave which is wholly positive.
The r.m.s. value is smaller in this case.)

For a RECTIFIED waveform the r.m.s. value may be deduced
as follows. For a half-wave rectified sine wave, the
r.m.s. value is half that of a complete sine wave:

$\frac{1}{2}$ × 0.7 × peak value

as shown in Fig.3.3a. For full-wave rectification, the
r.m.s. is the same as that for a complete sinusoidal waveform, namely 0.7 × peak value as shown in 3.3b. This is
because as far as r.m.s. calculations are concerned, a
negative-going half cycle is identical to a positive-going
half cycle.

Note that the d.c. or mean value of a waveform is merely
the average value over one cycle and thus bears no
relation to the r.m.s. value.

14

Vectorial or Phasor Representation

A sine wave may be represented by a vector or a phasor OA rotating anticlockwise at an angular velocity $\omega = 2\pi f$, where f is the frequency of the waveform (Fig.3.4). As the vector rotates, the height of its tip (point A) above the horizontal axis describes the sine wave shown. One complete revolution of the vector (360° or 2π) represents one complete cycle. Thus half a revolution (180° or π) represents half a cycle, and so on. The time axis of the waveform may, therefore, be used to represent the angle through which the vector has moved, as shown in the figure. The positive peak value is therefore at 90° (¼ cycle) and the negative peak is at 270° (¾ cycle).

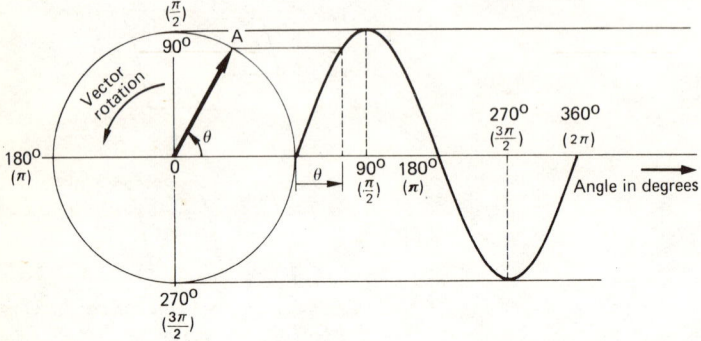

Fig.3.4 Vectorial representation of a sine wave

Now consider the two sine waves A and B, represented by vectors OA and OB respectively, in Fig.3.5. If both sine waves have the same frequency, then both OA and OB will rotate at equal angular velocity $\omega = 2\pi f$. This means that the angular difference θ between OA and OB does not change. OA is said to be *leading* OB by an angle of θ, or OB is *lagging* OA by an angle of θ. The sine waves are as shown in 3.5b.

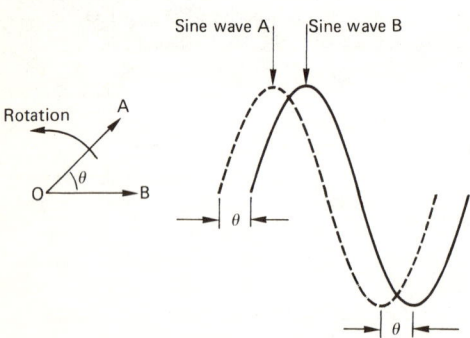

Fig.3.5 Phase difference.
OA leads OB (or OB
lags OA) by angle θ

Reactance

Inductors and capacitors offer a resistance to the flow of alternating current. This a.c. resistance is known as a reactance X and is measured in ohms. Reactance depends on the value of the inductor or the capacitor as well as the frequency of the a.c. waveform.

An inductor has an inductive reactance

$$X_L = 2\pi f L$$

where f is the frequency in Hz and L is the inductance in henries.

Since $\omega = 2\pi f$, then $X_L = \omega L$.

THE REACTANCE OF AN INDUCTOR INCREASES WITH FREQUENCY. (See Fig.3.6A.) For example a 10 mH inductor fed with a 1 kHz signal has a reactance

$$X_L = 2\pi \times 1 \times 10^3 \times 10 \times 10^{-3} = 2\pi \times 10 = 62.8\Omega$$

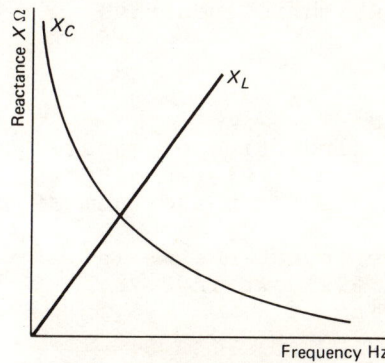

Fig.3.6A *Graph showing how the reactance of an inductor X_L and a capacitor X_C changes with frequency*

A capacitor has a capacitive reactance

$$X_C = \frac{1}{2\pi f C} = \frac{1}{\omega C}$$

where C is the capacitance in farads. THE REACTANCE OF A CAPACITOR DECREASES WITH INCREASING FREQUENCY. A 1 µF capacitor for example operating at 10 kHz has a reactance

$$X_C = \frac{1}{2\pi \times 10 \times 10^3 \times 1 \times 10^{-6}} = \frac{1}{2\pi \times 10^{-2}}$$

$$= \frac{10^2}{2\pi} = 15.9\Omega$$

A network consisting of a capacitive reactance X_C and an inductive reactance X_L has a total reactance given by the phasor sum of X_C and X_L. The two vectors X_C and X_L

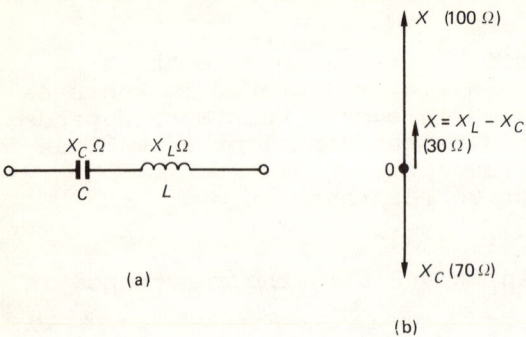

(a)

(b)

Fig.3.6B Vectorial sum of a capacitive X_C and an inductive X_L
reactance

are in anti-phase as shown in Fig.3.6B, i.e. they have a
phase difference of 180°. Hence the total reactance is
simply the difference between the values of X_C and X_L.
For example, given than $X_L = 100\Omega$ and $X_C = 70\Omega$, then the
total reactance $X = 100 - 70 = 30\Omega$ which is inductive
since X_L is greater than X_C.

Impedance

A network consisting of a reactance (inductive or capacitive) and a resistance has a total a.c. resistance known
as its impedance Z. The impedance Z is the phasor sum of
reactance X and resistance R.

For example, consider the reactive inductance X_L in
series with resistor R in Fig.3.7. As shown in 3.7b,
vector X_L *leads* vector R by 90° giving

$$\text{Impedance } Z = \sqrt{(X_L^2 + R^2)}$$

For $X_L = 400\Omega = 4 \times 10^2\Omega$ and $R = 300\Omega = 3 \times 10^2\Omega$,

$$Z = \sqrt{(16 \times 10^4 + 9 \times 10^4)} = \sqrt{(25 \times 10^4)} = (5 \times 10^2)$$

$$= 500\Omega$$

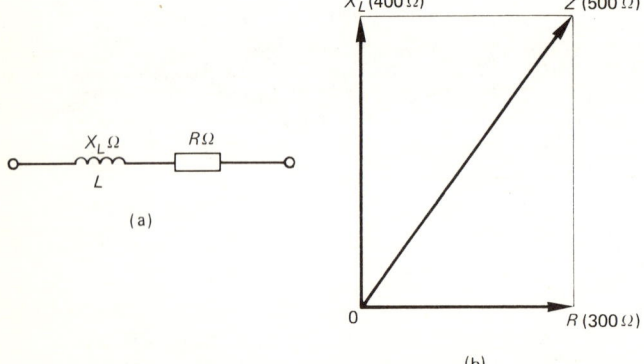

(a)

(b)

Fig.3.7 (a) Inductor L in series with resistor R
(b) Phasor representation of R, X_L and their phasor sum Z

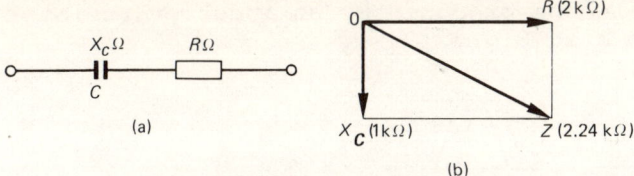

Fig.3.8 (a) Capacitor C in series with resistor R
(b) Phasor representation of R, X_C and their phasor sum Z

Fig.3.8 shows a capacitive reactance X_C in series with a resistor R. Vector X_C now *lags* vector R by 90° as shown in 3.8b, giving

Impedance $Z = \sqrt{(X_C^2 + R^2)}$

Given that $X_C = 1$ kΩ and $R = 2$ kΩ then

$Z = \sqrt{(1 + 4)} = \sqrt{5} = 2.24$ kΩ

For a network consisting of X_L, X_C and R the impedance Z is the vectorial sum of all three vectors.

Phase Shift

In many applications, a phase shift of a sine wave is required, e.g. in the phase shift oscillator (see p.142). Phase shift may be achieved by a simple capacitor-resistor (CR) network as shown in Fig.3.9. The output waveform voltage V_o is taken across the capacitor C. Compared with the input waveform the output displays the phase shift θ shown in 3.9b. The amount (or angle) of phase shift is determined by the values of C and R. The output has a smaller amplitude than the input due to the voltage drop across R.

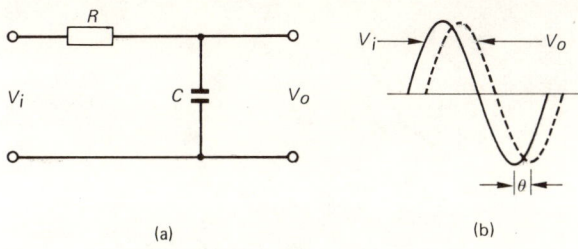

Fig.3.9 Phase shift produced by RC network

Phase shift may also be produced by an inductor-resistor (*LR*) network as shown in Fig.3.10.

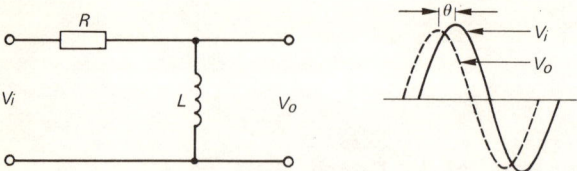

Fig.3.10 Phase shift produced by RL network

4 Resonance

Series Resonance

Consider the series network in Fig.4.1 with resistor R in series with inductor L and capacitor C. Resistor R may be an actual resistor or may represent the resistance of the inductor. The value of R is independent of frequency and stays constant at all frequencies. The reactances of L and C, on the other hand, are frequency-dependent. Thus as the frequency is increased from d.c. (zero Hz), the reactance of L increases while that of C decreases until at one frequency f_0, both reactances X_L and X_C become equal. The circuit is said to be at resonance, with f_0 as the frequency of resonance.

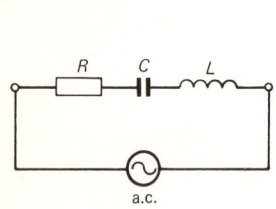

Fig.4.1 Series resonance

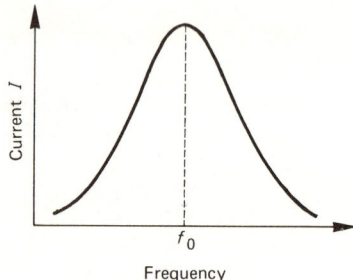

Fig.4.2 Response curve for a series resonant circuit

Being in anti-phase to each other X_L and X_C cancel each other, resulting in a purely resistive impedance $Z = R$. With Z at this minimum value, the current through the circuit is at maximum, giving the response curve shown in Fig.4.2. Because a series resonant circuit has a maximum current at resonance, it is also known as an ACCEPTOR circuit.

Parallel Resonance

Resonance may be obtained with the parallel circuit shown
in Fig.4.3. At one specified frequency determined by the
values of L and C the reactance of L cancels that of C
and produces resonance. The impedance is now at a maximum
giving maximum voltage. The circuit is thus known as a
REJECTOR, with a frequency response shown in Fig.4.4.

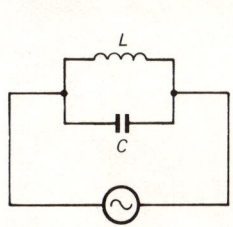

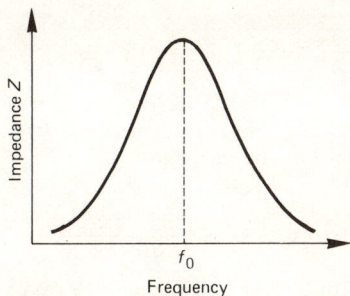

Fig.4.3 Parallel resonance

Fig.4.4 Response curve for a parallel
resonant circuit

   For both series and parallel resonance, the resonant
frequency is given as

$$f_0 = \frac{1}{2\pi\sqrt{(LC)}}$$

Given that L = 10 mH and C = 1 µF then

$$f_0 = \frac{1}{2\pi \times \sqrt{(10 \times 10^{-3} \times 1 \times 10^{-6})}}$$

$$= \frac{1}{2\pi\sqrt{10^{-8}}} = \frac{1}{2\pi \times 10^{-4}} = \frac{10^4}{2\pi} = 15.9\Omega$$

Note that the value of any resistance in the circuit does
not affect the frequency at which the resonance occurs.

Bandwidth and Selectivity

Resonant circuits are used mainly as tuned circuits
because of their selective frequency response. Normally,
parallel tuned circuits are used since they have a high
impedance and thus a high voltage output.
   The frequency response of a tuned circuit is shown in
Fig.4.5. The bandwidth is taken between the 3 dB points
as shown. The selectivity of a tuned circuit is a measure
of its ability to reject adjacent frequencies in favour
of the required tuned frequency. A more selective circuit
has a narrower bandwidth compared with a less selective
circuit having a wider bandwidth. Selectivity is measured
by its Q-factor which is defined as

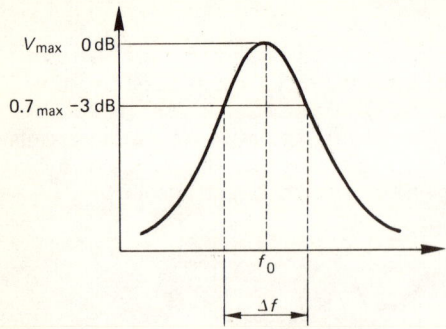

Fig.4.5 *Frequency response curve for a tuned circuit.*
Δf is the bandwidth

$$\text{Q-Factor} = \frac{\text{Resonant frequency}}{\text{Bandwidth}} = \frac{f_0}{\Delta f} \quad \text{(Fig.4.5)}$$

Hence a highly selective circuit has a high Q-factor; a less selective circuit has a low Q-factor.

Damping

Where a wide bandwidth is required, say for a TV receiver, low-Q tuned circuits are needed in the i.f. stage. The

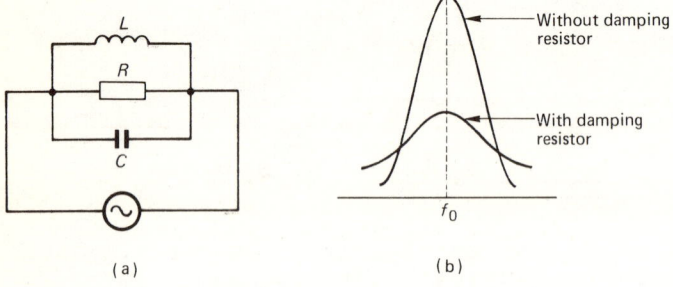

Fig.4.6 *Damping effect on a tuned circuit*

low Q-factor may be achieved by shunting the tuned circuit with a resistor R as shown in Fig.4.6a. The resistor has what is known as a damping effect on the response curve, making the circuit less selective, as shown in 4.6b.

Damped Oscillation

Consider a resonant circuit tuned to a frequency of 1 kHz. Since the circuit rejects all other frequencies, the only output that may appear across it will be in the form of a 1 kHz sine wave. Such an output may be produced by feeding the circuit with a.c. energy which makes it oscillate sinusoidally. This a.c. energy may be in the

form of a sine wave at the resonant frequency or a complex
waveform having a harmonic at the resonant frequency.

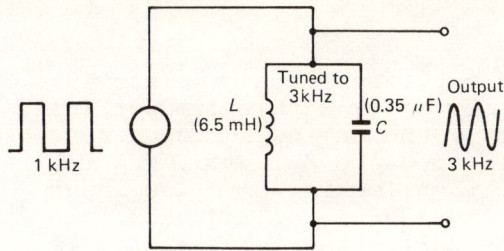

*Fig.4.7 A tuned circuit oscillating at the
        third harmonic of the input*

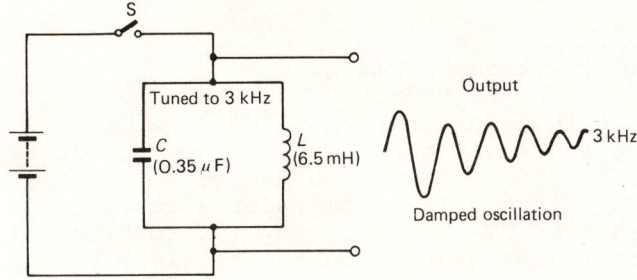

*Fig.4.8 Damped oscillation produced by a tuned
        circuit fed with a step waveform*

Fig.4.7 shows a resonant circuit tuned to 3 kHz fed with
a 1 kHz square wave. The tuned circuit will oscillate to
the third harmonic of the square wave (i.e. 3 kHz) producing sustained oscillations as shown. Fig.4.8 shows the
same tuned circuit, this time fed with a step voltage
generated by closing switch S. The step voltage contains
an infinite number of harmonics which energises the tuned
circuit into oscillating at its 3 kHz resonant frequency.
However, the oscillations are damped, i.e. dying out,
since the a.c. energy produced by the step voltage is
gradually lost due to the small losses in L and C as well
as in the resistance of the connecting wires, etc. After
a while the oscillations stop completely. Another set of
damped oscillations may be produced by opening switch S.
    Damped oscillation occurs whenever the energy fed into
the resonant circuit is not high enough or frequent
enough to overcome the losses in the circuit.

5 Matching

Input and Output Impedance

Any device such as a simple amplifier or a complete TV receiver may be represented by a box with two input leads and two output leads (Fig.5.1) and having an input impedance Z_{in} and an output impedance Z_O. Now, Z_{in} is the impedance of the device when looking into the input terminals AB, and Z_O is the impedance looking back into the output terminals CD. As far as the input signal is

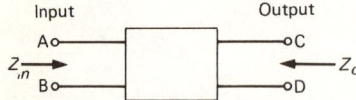

Fig.5.1 Input and output impedances Z_{in} and Z_O

concerned, the device is a load impedance of Z_{in}. As far as the next stage or a load is concerned, the device looks like a generator with an internal impedance of Z_O.

In many cases, e.g. in power amplifiers or a receiving aerial feeding a radio or TV receiver, we wish to transfer power from one stage to the other. In these cases it is important that the maximum possible power is transferred. This is the purpose of impedance matching or matching. Matching or maximum power transfer occurs when the output impedance of one stage is the same as the input impedance of the succeeding stage (Fig.5.2).

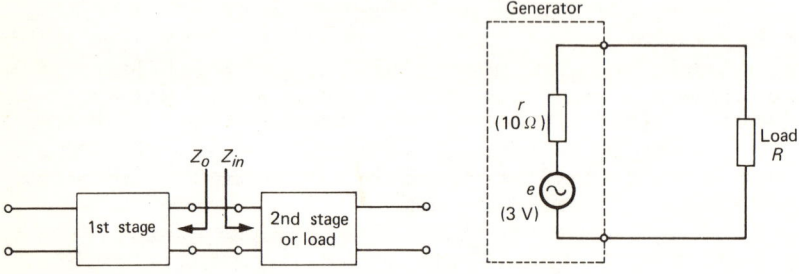

Fig.5.2 For matching $Z_O = Z_{in}$ Fig.5.3

As an illustration, consider the circuit in Fig.5.3 where a generator having an internal impedance of $r = 10\Omega$ and an e.m.f. $e = 3$ V is used to feed a load resistance R. Maximum power transfer from the generator to the load takes place when load R = internal (or output) resistance of the generator.

For $R = 10\Omega$, voltage across R, $V_R = 3/2 = 1.5$ V. Hence output power or power transferred into the load is

$$\frac{V^2}{R} = \frac{1.5 \times 1.5}{10} = \frac{2.25}{10} = 0.225 \text{ W} = 225 \text{ mW}$$

Note that the power dissipated into the load is the same as that dissipated in the internal resistance r when the load is correctly matched.

For any other value of load R, the output power will be smaller than 225 mW. For $R = 20\Omega$

$$V_R = \frac{3}{10 + 20} \times 20 = \frac{3 \times 20}{30} = 2 \text{ V}$$

giving an output power $= \frac{V^2}{R} = \frac{2 \times 2}{20} = \frac{4}{20} = 0.2 \text{ W} = 200 \text{ mW}.$

Similarly if load R were reduced in value to say $10/2$ or 5Ω, then output power will be reduced to 200 mW.

Examples

Example (1) Fig.5.4 shows an amplifier feeding an 8Ω loudspeaker in series with a 7Ω resistor.
(a) State the purpose of R.
(b) What is the main disadvantage of using this method?
Solution (a) The purpose of R is to ensure that the amplifier produces maximum power output.
(b) The disadvantage is that almost half of the output power is lost in the resistor.

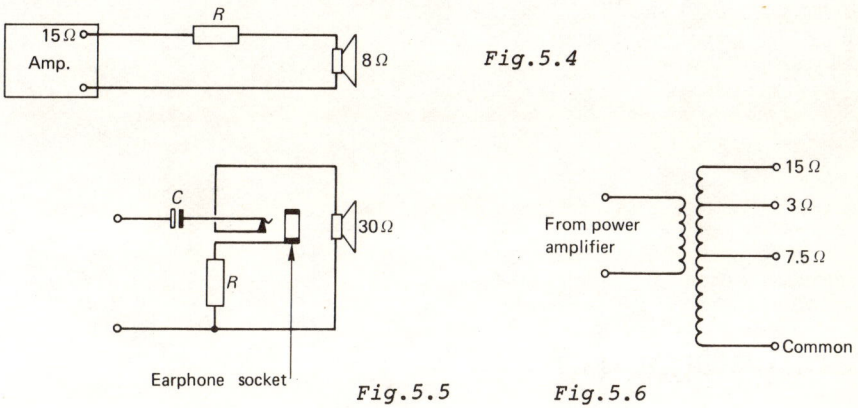

Fig.5.4

Fig.5.5   Fig.5.6

Example (2) Fig.5.5 shows a method used to provide a facility for earphone connection. State the value of R if the earphone has an impedance of 8Ω.
Solution $R = 30 - 8 = 22\Omega$.

Example (3) Refer to Fig.5.6. How should two 15Ω speakers be connected to the output transformer shown?

Solution For matching to take place, output impedance presented by the transformer must be equal to the load resistance. This can be done by connecting the two speakers in parallel, thus giving a total load of 7.5Ω across the common connection and the transformer tapping labelled 7.5Ω.

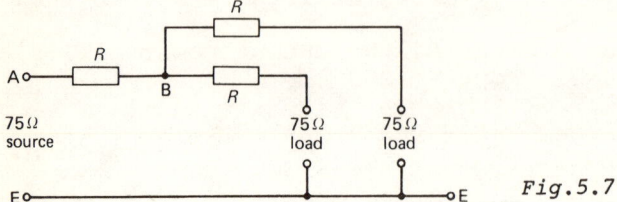

Fig.5.7

Example (4) Fig.5.7 shows a signal splitter network where an extra load is matched to the same source. Calculate the value of R.

Solution For matching the total resistance R_{AE} looking into AE must be 75Ω. The total resistance R_{BE} between B and E (across the two load branches) is

$$\frac{1}{2}(R + 75) = \frac{R}{2} + \frac{75}{2}$$

Total resistance across AE is

$$R_{AE} = R + R_{BE} = R + \frac{R}{2} + \frac{75}{2} = \frac{3R}{2} + \frac{75}{2}$$

But for matching, $R_{AE} = 75\Omega$. Therefore

$$75 = \frac{3R}{2} + \frac{75}{2}$$

i.e. $\frac{3R}{2} = 75 - \frac{75}{2} = \frac{75}{2}$

$$R = \frac{75}{3} = 25\Omega$$

6 Electromagnetism

Revision

When a current flows through a coil, a magnetic field is created in the vicinity of the coil. The strength of the magnetic field is determined by the number of turns of the coil as well as the current passing through it. A coil can store energy within its magnetic field. If a coil is placed inside a *changing* magnetic field, an induced e.m.f.

is produced across it. This fact is used in the TRANS-
FORMER. If a transformer has a turns ratio of

$$\frac{\text{Primary turns}}{\text{Secondary turns}} = \frac{T_1}{T_2} = n$$

then

$$\frac{V_1}{V_2} = \frac{T_1}{T_2} = n \quad \text{and} \quad \frac{I_1}{I_2} = \frac{T_2}{T_1} = \frac{1}{n}$$

Transformer Efficiency

It is assumed that a transformer is 100% efficient, i.e.
having no power loss. Hence

Input power $I_1 \times V_1$ = Output power $I_2 \times V_2$

In practice, transformers have efficiencies ranging from
96 to 99%.

Impedance Ratio

Consider the transformer shown in Fig.6.1 with a load
resistor r_2 connected across its secondary. Looking into
the primary, an input resistance r_1 is observed which is

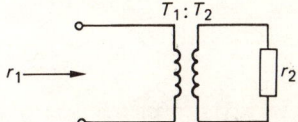

Fig.6.1 Impedance ratio $r_1/r_2 = T_1{}^2/T_2{}^2 = n^2$

the equivalent value of r_2 when transferred to the primary.
The ratio r_1/r_2 is known as the resistance or impedance
ratio of the transformer. This ratio may be derived as
follows. Since $r_1 = V_1/I_1$ and $r_2 = V_2/I_2$, then

$$\frac{r_1}{r_2} = \frac{V_1}{I_1} \div \frac{V_2}{I_2} = \frac{V_1}{I_1} \times \frac{I_2}{V_2} = \frac{V_1}{V_2} \times \frac{I_2}{I_1}$$

But $V_1/V_2 = T_1/T_2 = n$ and $I_2/I_1 = T_1/T_2 = n$, therefore

$$\frac{r_1}{r_2} = n \times n = n^2$$

For example, if $r_2 = 100\Omega$ and the turns ratio $T_1/T_2 = n$
= 2:1, then looking into the primary winding, the transformer will be seen as a resistor $r_1 = 100\Omega \times 2^2 =$
$100 \times 4 = 400\Omega$.

Transformer Matching

To ensure matching between two stages or between a load
and an output stage in a circuit, a transformer is some-

times used as the *coupling device*. Fig.6.2 shows an a.f.
power amplifier feeding into a loudspeaker load. Suppose
that a 10Ω loudspeaker is used with the amplifier having
an output impedance of 1000Ω. For matching to occur, the

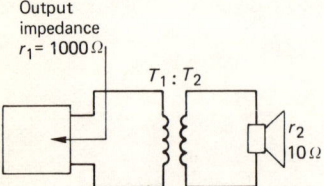

Fig.6.2 Matching transformer. Turns ratio 10 : 1

turns ratio $T_1/T_2 = n$ must be of such a value as to
present 1000Ω to the amplifier.

$$n^2 = \frac{r_1}{r_2} = \frac{1000}{10} = 100 \quad \text{Hence} \quad n = \sqrt{100} = 10.$$

In other words, a step-down transformer must be used
having a turns ratio of 10:1.

Note that in all cases where the amplifier is feeding a
low impedance load (which is the majority of cases in
practice), the matching transformer used must have a
STEP-DOWN turns ratio.

Auto-transformer

A transformer may be made with a single winding employing
one (or more) tappings as shown in Fig.6.3. T_a is the
number of turns of the primary and T_b is the number of
turns of the secondary. The voltage, current, impedance,
and turns ratios have the same relationship as for an
ordinary transformer.

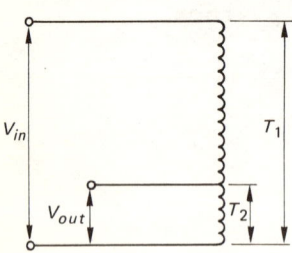

Fig.6.3 The auto-transformer

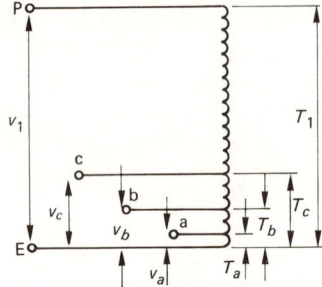

*Fig.6.4 An auto-transformer with
multiple tappings*

Fig.6.4 shows another single winding transformer employing more than one tapping. The voltage, current, and
impedance ratios are once again determined by the turns
ratio, e.g. $v_1/v_a = T_1/T_a$ and $v_1/v_b = T_1/T_b$ and so on.

The Centre Tapped Transformer

Fig.6.5 shows a transformer with a tap at the centre of its secondary. Two output voltages v_a and v_b are produced across each half of the secondary winding. The ratio of the input (primary) voltage to the secondary voltages is given by the turns ratio whereby

$$\frac{v_1}{v_a} = \frac{T_1}{T_a} \quad \text{and} \quad \frac{v_1}{v_b} = \frac{T_1}{T_b}$$

where T_1, T_a and T_b are the number of turns of the primary, secondary a and secondary b respectively. Since the tapping is at the centre, v_a and v_b have equal amplitude. If the centre is earthed, as shown in Fig.6.5, the two output waveforms are then in anti-phase to each other.

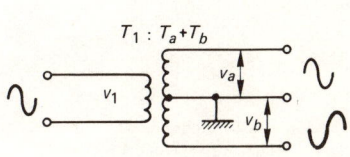

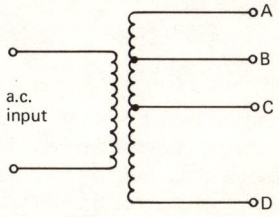

Fig.6.5 The centre tapped
        transformer

Fig.6.6

Example Refer to Fig.6.6. (a) Calculate the voltage between transformer tappings B and C. (b) If there are 30 turns between tappings A and B, how many turns will the whole secondary winding have?
Solution
(a) Voltage between B and C = $v_{BC} = v_{AD} - v_{AB} - v_{CD}$

$$= 36 \text{ V} - 6 \text{ V} - 12 \text{ V} = 12 \text{ V}.$$

(b) $\dfrac{v_{AB}}{v_{AD}} = \dfrac{\text{Number of turns between A and B}}{\text{Number of turns between A and D}}$

Hence $\dfrac{6 \text{ V}}{36 \text{ V}} = \dfrac{30}{T_{AD}}$ or $T_{AD} = 30 \times \dfrac{6}{36} = 5$ turns

The Magnetic Circuit

It is accepted that, in a magnetic circuit, the magnetic flux (or magnetic field) measured in Teslas is created by a force known as the magnetomotive force m.m.f. A magnetic circuit is usually compared with an electric circuit, where the flux corresponds to the current and the magnetomotive force m.m.f. corresponds to the electromotive force e.m.f. Just as an electric circuit is said to have a resistance, so a magnetic circuit is said to have a RELUCTANCE S which corresponds to the reciprocal of R, i.e. $1/R$. For example a magnetic material such as soft iron has a high reluctance, i.e. low resistance to magnetic flux.

Permeability

The permeability of a material is a measure of the ease with which it may be magnetised. For example, soft iron and other electromagnetic materials such as ferrites have high permeability. These materials are used in transformers, inductors, relays, and in ferrite aerials. On the other hand, non-magnetic materials have low permeability. Magnetic alloys such as silicon steel have the ability to retain their magnetism and are therefore used as permanent magnets in loudspeakers, moving coil instruments, etc.

Screening

Consider the hollow cylinder, shown in Fig.6.7, placed inside a magnetic field. If the cylinder is made of a high reluctance magnetic material, then the magnetic field will concentrate through the cylinder as shown, with no magnetic flux passing through the space inside the cylinder. Consequently a body placed in this space would be screened or shielded from the magnetic field around it. Such screening, known as magnetic screening, is used to protect devices such as a cathode ray tube, a moving coil meter, or a loudspeaker from outside magnetic fields.

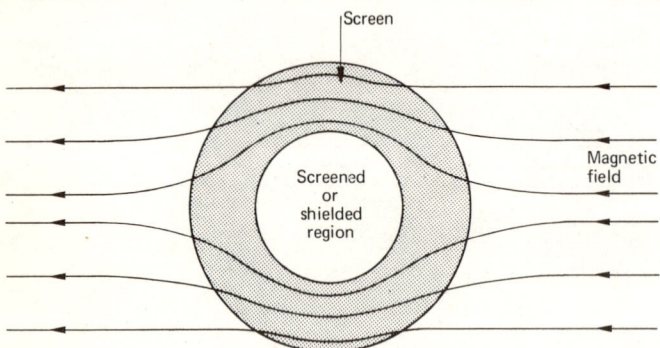

Fig.6.7 Magnetic screening

In transformers, another type of screening known as electrostatic or electric screening is sometimes used. A screen or shield of thin copper sheet is placed between the primary and secondary windings as shown in Fig.6.8. By earthing the screen, the effect of the capacitance that exists between the two windings due to their different

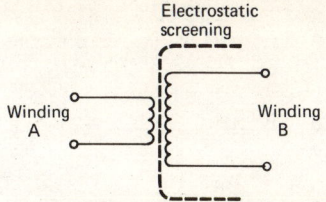

Fig.6.8 Electrostatic screening as used in a transformer

potentials is minimized. Electrostatic screening is also used in coaxial cables and tetrode and pentode valves or whenever conductors having different potentials are placed in close proximity.

7 The Decibel

The decibel dB is a logarithmic unit for measuring gain. Initially it was used to measure power gain. The decibel is found to be a more suitable unit to use than the simple gain ratio because it provides a convenient number for the gain even if the power gain ratio is very high. It is also more convenient to use in a.f. applications because the human ear responds to sound intensities on a logarithmic scale.

Input p_{in} ▷ Output p_o *Fig.7.1*

For the amplifier shown in Fig.7.1 where p_{in} and p_o are the input and output powers, the power gain G_p in dB is given as

$$G_p = 10 \log \frac{\text{Output power}}{\text{Input power}} = \log \frac{p_o}{p_{in}}$$

Since p_o/p_{in} = power gain, then

dB gain = 10 log (power gain)

It should be noted that the decibel is a unit or power ratio and not a measure of the power level.

The dB gain may be positive or negative. A positive dB represents a gain ratio greater than 1 (p_o greater than p_{in}), while a negative dB represents a loss or a gain ratio smaller than 1 (p_o smaller than p_{in}).

To calculate the dB gain using the formula given above requires a knowledge of logarithms. It is therefore useful to memorise the basic conversions given in Table 7.1 which as we shall see later may be used to calculate most of the conversions that are likely to crop up.

It should be noted that (*a*) for a gain of 1 the dB gain
is zero and (*b*) that a 10 time increase in gain (e.g. from
100 to 1000) the dB gain increases by 10 dB (from 20 to
30). For a loss, the dB gain is negative, as stated
earlier. For example, a gain of ½ = -3 dB and 1/10 =
-10 dB and so on.

TABLE 7.1

dB gain	Power gain
O	1
3	2
10	10
20	100 (10^2)
30	1000 (10^3)

TABLE 7.2

dB gain	Voltage or current gain
O	1
3	1.4
6	2
20	10
40	100
60	1000

Examples
(1) If an amplifier has a gain of 3 dB, and an input of
10 mW then the output may be calculated as follows:

3 dB = a power gain of 2

Hence, Output = 2 × input = 2 × 10 mW = 20 mW

Gain = 10 dB

o—[▷]—o100 mW   *Fig.7.2*

(2) Refer to Fig.7.2. To find the input power:

10 dB = a power gain of 10

Hence, Input power = $\dfrac{\text{Output power}}{10}$ = $\dfrac{100 \text{ mW}}{10}$ = 10 mW

Voltage and Current Gain

From the dB gain of a device, both the voltage and current
gains can be calculated as follows:

$$\text{Gain in dB} = 20 \log \frac{v_o}{v_{in}} = 20 \log \frac{I_o}{I_{in}}$$

where v_o and I_o are the output and v_{in} and I_{in} are the
input voltage and current respectively.

From Table 7.2 -3 dB = a voltage gain of 1/1.4 = 0.7 or
70%. For example, if a 3 dB attenuator (i.e. having a
dB gain of -3 dB) has an input voltage of 10 mV then the
output is 0.7 × 10 = 7 mV.

Multi-stage Gain

One further advantage of the dB scale is that the dB gain
of an amplifier having more than one stage is given as
the sum of the dB gain of each stage. Fig.7.3 shows a
two-stage amplifier. Stage 1 has a gain of 10 dB (i.e. a
power gain of 10) and stage 2 has a gain of 3 dB (or a
power gain of 2). The overall gain is 10 dB + 3 dB =
13 dB. The overall gain ratio on the other hand is given
as the product of the gain of each stage, namely 10 × 2 = 20.

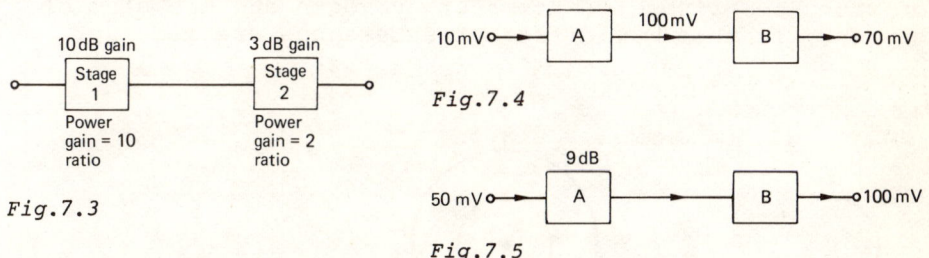

Fig.7.3

Fig.7.4

Fig.7.5

Example Refer to Fig.7.4.

   Stage A gain = 100/10 = 10, or 20 dB

   Stage B gain = 70/100 = 0.7, or - 3 dB.

   Overall gain = 70/10 = 7.   Overall dB gain = 20 - 3
                                                = 17 dB.

Conversion from dB to ratios or vice versa for values
other than those given in the two tables may be worked
out by sub-division into suitable stages. For example,
13 dB may be divided into 10 dB + 3 dB, and 10 dB is a
ratio of 10 and 3 dB is a ratio of 2. Hence the total
equivalent ratio is 10 × 2 = 20 (refer to Fig.7.3).

Examples
(1) If the input to a device is 1 mW and its output is
1 W what is the gain in dB?

Solution Power gain = $\dfrac{1\ W}{1\ mW}$ = $\dfrac{1000\ mW}{1\ mW}$ = 100

   A power gain of 100 = 20 dB.

(2) Refer to Fig.7.5. Find (*a*) the overall gain, (*b*) the
overall dB gain, (*c*) the gain of stage B in dB and as a
ratio.

Solution

(*a*) Overall gain = $\dfrac{v_o}{v_{in}}$ = $\dfrac{100\ mV}{50\ mV}$ = 2

(*b*) Voltage gain of 2 = 6 dB.

(*c*) Gain stage B = 6 dB - 9 dB = - 3 dB.

   Stage B gain as a ratio is 1/1.4 = 0.7.

8 The Semiconductor Diode

Conductors and Insulators

All substances are made up of one or more chemical elements such as oxygen, sulphur, etc. The smallest part of an element is the atom. Atoms of different elements may combine to form a molecule of a substance, e.g. a molecule of water comprises two atoms of hydrogen and one atom of oxygen. In this way different substances are made up.

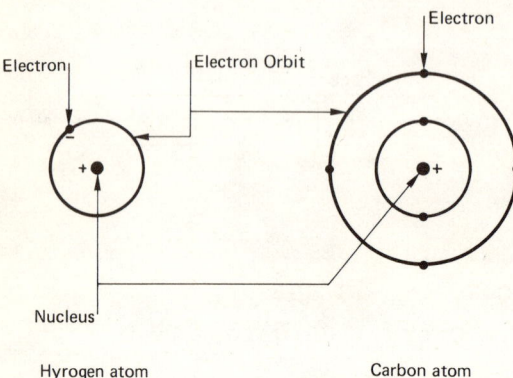

Fig.8.1 Atoms consist of negative electrons rotating round a positive nucleus

The atom itself is composed of smaller particles known as electrons revolving around a central nucleus containing one or more protons (Fig.8.1). The electrons which are negatively charged are attracted to the positive protons and continuously rotate in orbits or shells round the nucleus. The number of electrons in an atom is exactly balanced by the number of protons.

Atoms of different elements are differentiated one from the other by the number of electrons it contains, e.g. a hydrogen atom has one electron while that of carbon has six. For electrical conduction to occur, electrons which are loosely connected to the nucleus (known as free electrons) leave their orbit when attracted by a positive potential, and begin to travel along forming a flow of electrons, or current. A good conductor has a large number of "free electrons" while an insulator has extremely few "free electrons".

Semiconductors

Semiconductors have their atoms grouped together in a
regular pattern called a "crystal lattice". They are not
good conductors (hence their name) since they have few
"free electrons". The number of "free electrons" increases as the temperature goes up, which leads to
improved conductivity. These "free electrons" are known
as minority carriers.

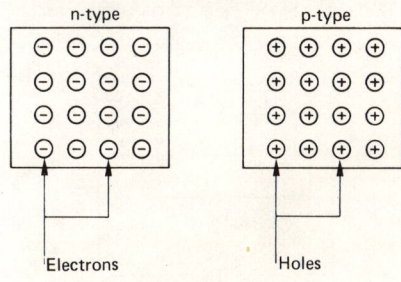

n-type   p-type

Electrons   Holes

Fig.8.2 n-type and p-type
semiconductors

Conductivity may also be improved by adding controlled
amounts of "impurities". Impurities such as arsenic
atoms introduce extra electrons in the lattice structure
producing the n-type semiconductor. Such atoms are known
as *donor* atoms. On the other hand impurity atoms known
as *acceptor* atoms, for instance aluminium atoms, introduce a "shortage of electrons" referred to as holes and
producing the p-type semiconductor (Fig.8.2). The
electrons and holes produced by the injection of impurities are known as the majority carriers.

The Junction Diode

When a p-type semiconductor is joined to an n-type semiconductor (Fig.8.3), by a process known as *diffusion*,
electrons from the n-region begin to cross over to fill
the holes in the p-region. This movement continues until
a neutral zone known as the *depletion layer* is established
on either side of the pn junction. This depletion layer
provides a potential barrier preventing further movement
by electrons across the junction.

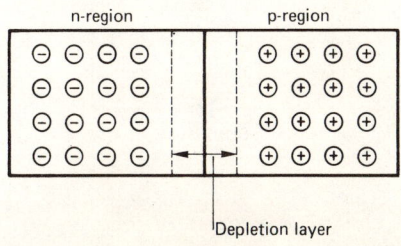

n-region   p-region

Depletion layer

Fig.8.3 The pn junction diode

Electrons wishing to cross over must now have sufficient energy to overcome the potential barrier. Such energy may be provided by an external e.m.f. The height of the potential barrier depends on the semiconductor used, namely 0.3 V for Germanium (Ge) and 0.6 V for Silicon (Si).

Diode Characteristics

When the diode is reverse biased (Fig.8.4), electrons in the n-region are attracted to the positive electrode of the bias voltage, while the holes in the p-region are

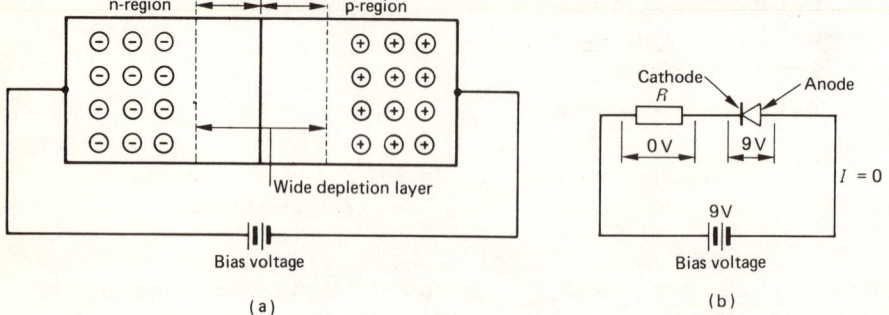

Fig.8.4 Reverse biased diode. (a) shows the increase in the width of the depletion layer

attracted to the negative electrode. The depletion layer is thus widened, producing a higher potential barrier resisting further the flow of electrons.

A forward bias, however, removes the depletion layer as shown in Fig.8.5, allowing the electrons to cross over, with current due to majority carriers flowing freely. The diode however maintains a constant voltage drop across it known as the forward voltage drop (0.3 V for Ge and 0.6 V for Si diodes).

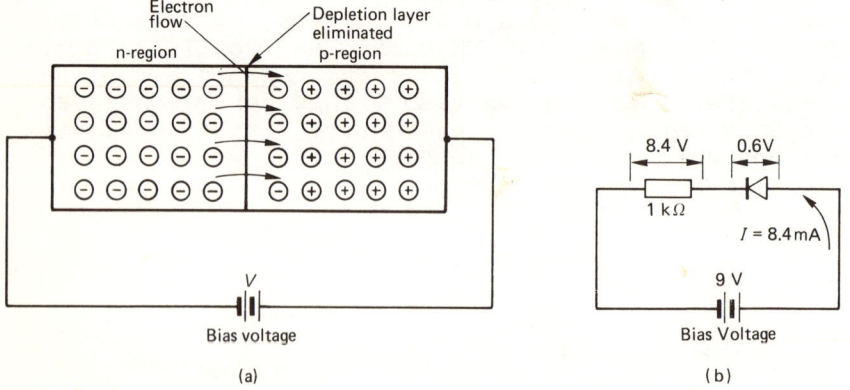

Fig.8.5 Forward biased diode. (a) shows the elimination of the depletion layer

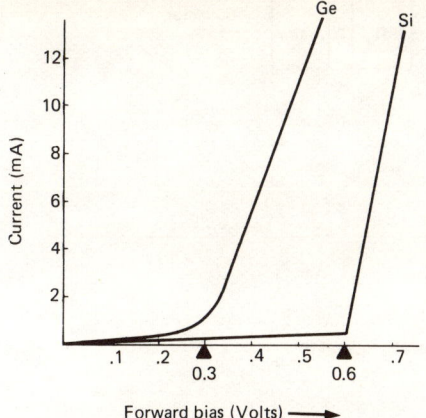

Fig.8.6 *Forward characteristics
of germanium and
silicon diodes*

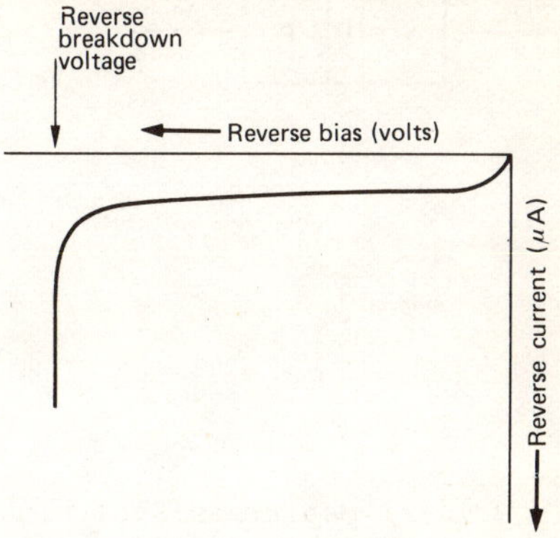

Fig.8.7 *Reverse characteristics
of a junction diode*

The junction diode forward characteristics are as shown
in Fig.8.6. Note that as soon as the bias voltage over-
comes the potential barrier of the diode, a large current
begins to flow. A *very small* increase in the bias voltage
produces a *very large* increase in the current through the
diode. For bias voltages below the required forward
voltage drop, a small leakage current (in microamps)
flows which is usually neglected.

The reverse bias characteristics are as shown in Fig.
8.7. An extremely small current passes through the diode
when in reverse bias, due to minority carriers. This
reverse current is almost constant up to a maximum voltage
known as the reverse breakdown voltage. If a higher
voltage is applied, breakdown occurs and the reverse
current increases rapidly causing damage to the diode.
Care must always be taken to ensure that diodes are not
subjected to reverse voltages higher than the breakdown
voltage specified by the manufacturer. Germanium diodes
have higher leakage currents and therefore lower reverse
resistance than silicon diodes.

9 Transistors

The transistor is a semiconductor device made up of two
pn junctions as shown in Fig.9.1. There are three terminals to the transistor, known as the emitter, base, and
collector. Transistors may be of two types: the pnp
shown in (a) and the npn shown in (b). The principle of

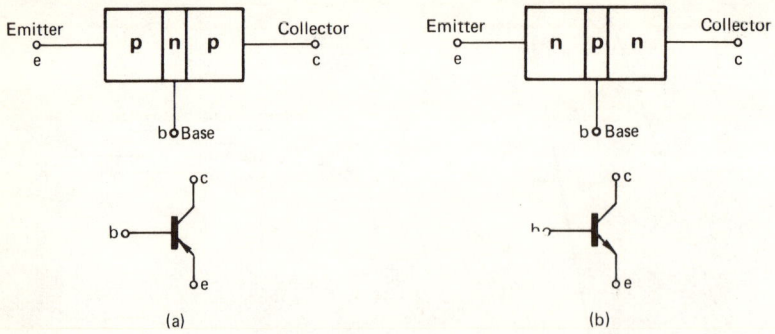

Fig.9.1 (a) pnp transistor with its symbol
(b) npn transistor with its symbol

operation for both types is the same, the only difference
being in the d.c. bias conditions.

Consider the npn transistor shown in Fig.9.2. The base-
emitter (b-e) junction is forward biased by voltage V_2.
Electrons from the emitter (I_e) will, therefore, flow
across the junction from emitter to base. It is the
normal forward current for a forward bias pn junction. As
soon as the electrons cross over into the base, they are
attracted by the positive potential of the collector. By
making the base region very thin it is possible to have
almost all of these electrons move across the base region
towards the collector. Only a very small number of
electrons is collected by the base forming the base
current I_b. In fact 95% or more of the emitter current I_e
is now collected by the collector to form the collector
current I_c. Therefore,

$$I_e = I_c + I_b$$

Since base current I_b is very small (in μA) it is usually
neglected. I_c and I_e are assumed to be equal and are
generally referred to as the transistor current.

It will be noticed that the base-collector (b-c)
junction is reversed biased by voltage V_1. This is
essential otherwise the electrons will not be attracted
towards the collector.

For a pnp transistor the polarities of the d.c. supplies
must be reversed as shown in Fig.9.3. In this case the
transistor current is a movement of holes from emitter to
collector.

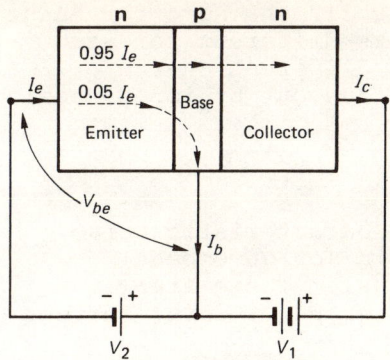

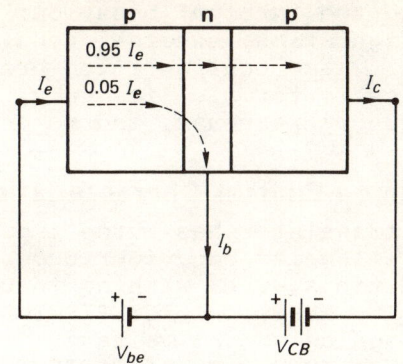

Fig.9.3 *Bias voltages for pnp*
        transistor

Fig.9.2 *Bias voltages for npn*
        transistor

Transistor Configurations

There are three possible ways to connect a transistor when used in a circuit.

(1) The COMMON EMITTER (CE) configuration where the input signal is fed between the base and emitter, while the output is taken between collector and emitter (Fig. 9.4). This is the most widely used configuration due to its flexibility and high gain.

(2) The COMMON BASE (CB) configuration where the base is common to both input and output (Fig.9.5).

(3) the COMMON COLLECTOR (CC) configuration where the collector is common. This configuration is also known as the emitter follower since the output is taken at the emitter (Fig.9.6).

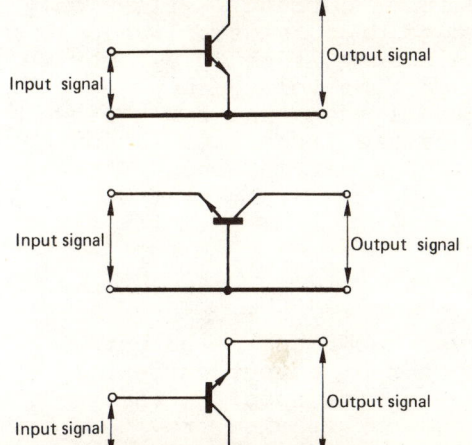

Fig.9.4 *Common Emitter (CE)*
        configuration

Fig.9.5 *Common Base (CB)*
        configuration

Fig.9.6 *Common Collector (CC)*
        configuration. (Note
        that the output is
        taken from the emitter)

While the internal behaviour is exactly the same for each configuration, external behaviour is different for each case. Each configuration presents different characteristics in terms of its gain, input and output impedances, frequency response, and so on.

Common Emitter Characteristic Curves

The transistor has three types of characteristic curve outlining its behaviour under static (or quiescent) conditions, i.e. with no input signal. These are:
(1) The INPUT CHARACTERISTICS, the graph of *input* current against *input* voltage.
(2) The OUTPUT CHARACTERISTICS, the graph of *output* current against *output* voltage.
(3) The TRANSFER CHARACTERISTICS, the graph of *output* current against *input* current.
The characteristics described below are for an npn trans-

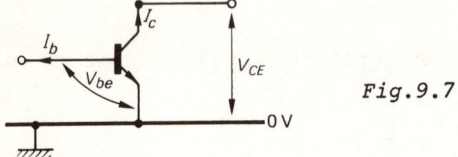

Fig.9.7

istor (Fig.9.7). For a pnp transistor the d.c. voltages should be negative.

Input Characteristics

Fig.9.8 shows the input characteristics for an npn transistor. It is the same as the characteristics of a forward biased pn junction since the input (b-e) is a forward biased junction. It will be noted that just as in the case of a junction diode, the input current I_b flows only when the required forward voltage drop is established across the b-e junction. Once that voltage (0.3 V for Ge and 0.6 V for Si) is established, the voltage between base and emitter V_{be} remains practically constant for a large increase of base current. Hence the transistor is considered as a current device, i.e. a varying input current with a constant input voltage.

Output Characteristics

Fig.9.9 shows a family of curves known as the output characteristics relating collector or output current I_c to collector or output voltage V_{CE} for specific values of base or input current I_b. These curves also provide a relationship between the input current on the one hand and output current and voltage on the other. For example,

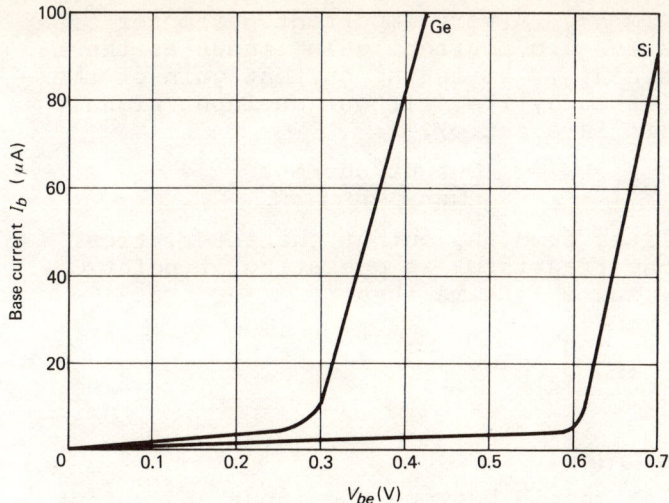

Fig.9.8 *Input characteristics of a transistor*

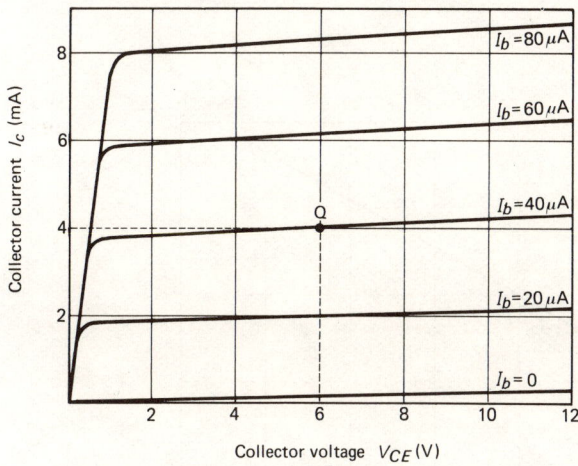

Fig.9.9 *Output characteristics of a transistor*

given a transistor with a base current of 40 μA and a
collector voltage of 6 V, then using the output characteristics in Fig.9.9 a collector current of 4 mA will be
flowing.

The curve I_b = 0 represents the state of the transistor
before it begins to conduct, i.e. at cut off when V_{be} is
below the required forward voltage drop. Theoretically,
transistor current is zero when I_b = 0; however, a small
leakage current does flow through to the collector.

STATIC CURRENT GAIN h_{fe} A very important parameter of any transistor is its d.c. current gain, known as the static current gain h_{fe}. It is the current gain of the transistor at quiescence, i.e. without an input signal. It has no units (it is a ratio).

$$\text{Static current gain } h_{fe} = \frac{\text{Output current}}{\text{Input current}} = \frac{I_c}{I_b}$$

h_{fe} can be calculated from the output characteristics. For example, if the transistor is operating at point Q where I_b = 40 µA, and I_c = 4 mA then

$$h_{fe} = \frac{I_c}{I_b} = \frac{4 \times 1000}{40} = 100$$

Transfer Characteristics

These are the relationship between the input and output currents (Fig.9.10). The static current gain may also be calculated using this characteristic curve. For example if Q is the operating point of the transistor then

$$h_{fe} = \frac{I_c}{I_b} = \frac{4 \times 1000}{40} = 100$$

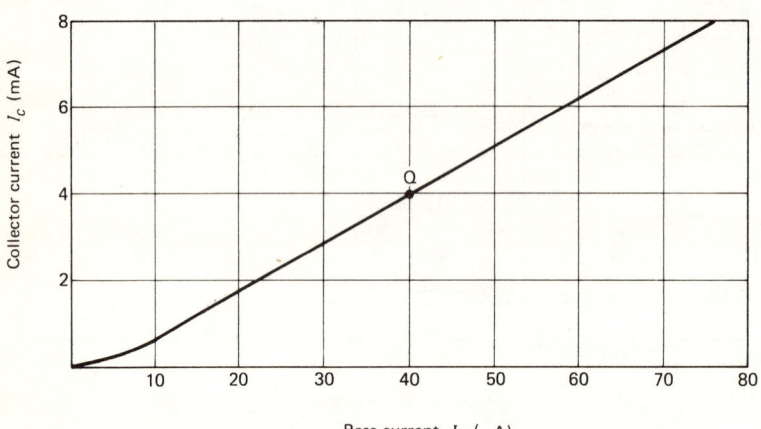

Fig.9.10 Transfer (or mutual) characteristics of a transistor

10 The Common Emitter Amplifier

Before the transistor can be used as an amplifier it must
have the correct d.c. bias, as shown in Fig.10.1a for an
npn transistor. It can be seen that the two voltages V_2
(providing the forward bias for the b-e junction) and V_1
(providing the reverse bias for the b-c junction) are

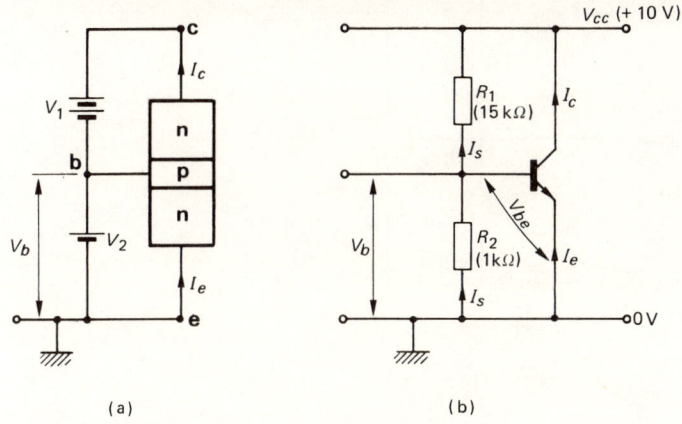

(a)   (b)

Fig.10.1 Base bias for npn transistor

connected in series. V_1 and V_2 can thus be substituted
by the potential divider R_1-R_2 shown in 10.1b. This way
only one d.c. supply voltage is required, namely V_{cc}.
The ratio of R_1 to R_2 is chosen so that the transistor is
given the required bias at the base.
   A standing current $I_s = V_{cc}/(R_1 \times R_2)$ flows through the
bias chain R_1-R_2 causing a drain on the supply voltage.
To reduce this standing current high values of R_1 and R_2
are used. However, as will be seen later, a very high
value of R_1 reduces the d.c. stability of the transistor
due to base current.
   The voltage at the base is the voltage between the base
and the O V line or chassis, i.e. the voltage across
resistor R_2.

Voltage at the base $V_b = V_{R_2} = \dfrac{V_{cc}}{R_1 + R_2} \times R_2.$

For example, given $V_{cc} = 10$ V, $R_1 = 15$ kΩ, $R_2 = 1$ kΩ, then

$$V_b = \frac{V_{cc}}{R_1 + R_2} \times R_2 = \frac{10 \text{ V}}{15 \text{ k}\Omega + 1 \text{ k}\Omega} \times 1 \text{ k}\Omega$$

$$= \frac{10}{16} \times 1 = 0.62 \text{ V}$$

By using different values of R_1, R_2 or both, the base voltage may be varied.

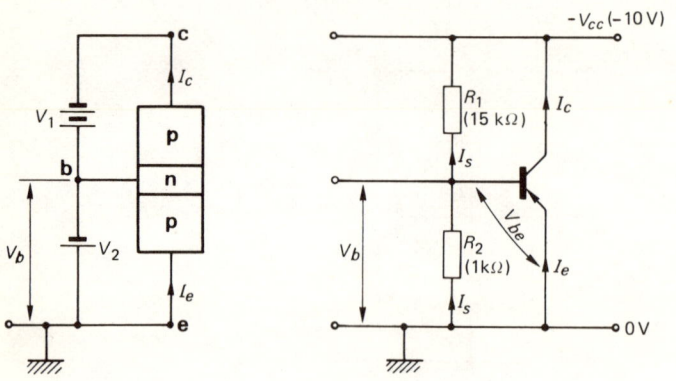

Fig.10.2 Base bias for pnp transistor

The same biasing arrangement may be used for a pnp transistor as shown in Fig.10.2. In this case the d.c. supply is negative, $-V_{cc}$. The current, as was explained earlier, may be considered to flow from the emitter to the collector as in the case of the npn transistor, only this time it is a movement of holes. The function of bias chain R_1-R_2 is the same as that for the npn transistor. The fact that the d.c. voltages are negative should be remembered but need not be included in calculation. Hence

$$V_b = \frac{V_{cc}}{R_1 + R_2} \times R_1 = \frac{10}{15 + 1} \times 1 = \frac{10}{16} = 0.62 \text{ V}$$

The voltage at the base is -0.62 V.

For a forward bias to be established across the b-e junction, the base voltage has to be "higher" than the emitter voltage - "higher", whether in the positive (npn) or in the negative (pnp) directions. In general, regardless of the type of transistor used, the base voltage is always higher than the emitter but lower than the collector.

As explained in the previous section, the current through the transistor is determined by the b-e forward bias V_{be}, i.e. the potential difference between base and emitter $V_{be} = V_b - V_e$. A variation in emitter or base voltage causes a change in transistor current. In the transistor circuit under consideration the emitter being at chassis potential, only the base can change.

$$V_{be} = 0.62 - 0 = 0.62 \text{ V}$$

For example, if the base voltage V_b goes away from the emitter (more positive for npn or more negative for pnp transistors), V_{be} increases causing the transistor current to go up. On the other hand if V_b goes towards the emitter, V_{be} decreases and the transistor current is reduced.

Collector or Load Resistor

In order to produce an output at the collector, a load resistor R_3 (also known as the collector resistor) is

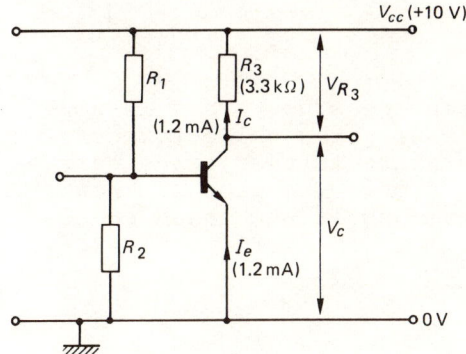

Fig.10.3 *The load resistor* R_3

added as shown in Fig.10.3. The collector current I_c flows through collector resistor R_3 developing a voltage across it. Hence

$$V_{R_3} = I_c \times R_3$$

Since all voltages are measured with respect to chassis, the collector voltage V_c is the p.d. between the collector and the chassis. As can be seen from the circuit,

Supply voltage $V_{cc} = V_{R_3} + V_c$

Hence $V_c = V_{cc} - V_{R_3}$

Taking the typical values shown where $I_c = 1.2$ mA,

$$V_{R_3} = I_c \times R_3 = 1.2 \text{ mA} \times 3.3 \text{ k}\Omega = 4 \text{ V (approx.)}$$

$$V_c = V_{cc} - V_{R_3} = 10 - 4 = 6 \text{ V}$$

Thermal Runaway

It has been mentioned in Chapter 8 that minority carriers form what is known as leakage current across a reverse biased junction. A leakage current I_{CBO} will, therefore,

flow from collector to base as shown in Fig.10.4. It is
amplified in the same way as the input (or base) current
by a factor of h_{fe}. As the temperature of the transistor

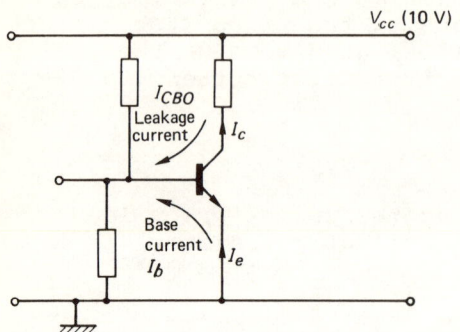

Fig.10.4 Leakage current I_{CBO}

increases, so the leakage current goes up. This is amplified by the transistor, increasing the collector current
which in turn raises the temperature further, causing the
leakage current to go up, and so on. The process known
as thermal runaway is cumulative and if unchecked could
cause damage to the transistor.

D.C. Stabilisation

The effect of leakage current in the CE amplifier is to
cause instability in d.c. or static operating conditions.
This can be overcome by the addition of emitter resistor
R_4 as shown in Fig.10.5. The emitter voltage is now equal
to the volt drop produced by the emitter current I_e flowing
through emitter resistor R_4. Hence $V_e = I_e \times R_4$. D.C.
stabilisation takes place as follows.

If I_c and I_e increase due to an increase in leakage
current, the emitter voltage V_e increases with it. Since
$V_{be} = V_b - V_e$ then an increase in V_e produces a decrease
in V_{be}. This reduces the base current and restores I_c and
I_e to their previous values. The emitter resistor R_4
provides a negative feedback loop giving stability to the
static conditions of the amplifier. Using the typical
values given in Fig.10.5 and assuming emitter current
I_e = 1.2 mA, then $V_e = I_e \times R_4$ = 1.2 mA × 1 kΩ = 1.2 V.

$$V_b = \frac{V_{cc}}{R_1 + R_2} \times R_2 = \frac{10}{15 \text{ k}\Omega + 3.3 \text{ k}\Omega} \times 3.3 \text{ k}\Omega$$

$$= \frac{10}{18.3} \times 3.3 = 1.8 \text{ V}$$

$$V_{be} = V_b - V_e = 1.8 - 1.2 = 0.6 \text{ V}$$

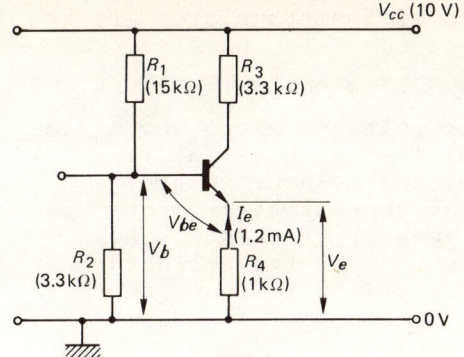

Fig.10.5 npn common emitter
amplifier with
emitter resistor R_4

Using pnp Transistors

Fig.10.6 shows a pnp transistor amplifier. Assuming the
transistor to be silicon, its current, emitter, base and

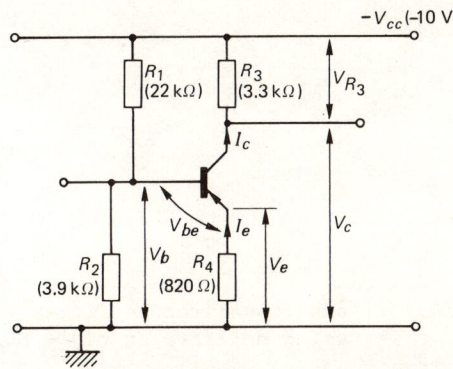

Fig.10.6 pnp common emitter
amplifier

collector voltages can be calculated as follows:

$$V_b = \frac{V_{cc}}{R_1 + R_2} \times R_2 = \frac{10}{22\ k\Omega + 3.9\ k\Omega} \times 3.9\ k\Omega$$

$$= \frac{10}{25.9} \times 3.9 = 1.5\ V$$

From $V_{be} = V_b - V_e$ we get $V_e = V_b - V_{be}$ but $V_{be} = 0.6\ V$
(silicon transistor) and $V_b = 1.5\ V$. Therefore

$$V_e = 1.5 - 0.6 = 0.9\ V$$

From $V_e = I_e \times R_4$, we get

$$I_e = \frac{V_e}{R_4} = \frac{0.9\ V}{820\ \Omega} = \frac{0.9\ V}{0.82\ k\Omega} = 1.1\ mA$$

$$I_c = I_e = 1.1\ mA$$

$$V_{R_3} = I_c \times R_3 = 1.1\ mA \times 3.3\ k\Omega = 3.6\ V$$

$$V_c = V_{cc} - V_{R_3} = 10 - 3.6 = 6.4\ V$$

The static operating conditions of the transistor are therefore

e -0.9 V, b - 1.5 V, c - 6.4 V with I_e = 1.1 mA

The emitter, base and collector voltages given above are typical for a single stage amplifier such as an i.f. amplifier or a driver. The emitter is approximately 0.1 × V_{cc}, with the collector at approximately 0.6 × V_{cc}. It can be seen that for both types of transistor the emitter voltage is lowest while that of the collector is highest, with the base at 0.6 V (for Si transistors) above the emitter.

The npn Transistor with Negative Supply

It is possible to connect an npn transistor using a negative d.c. supply voltage $-V_{cc}$ as shown in Fig.10.7. In this case the chassis is the positive rail. All

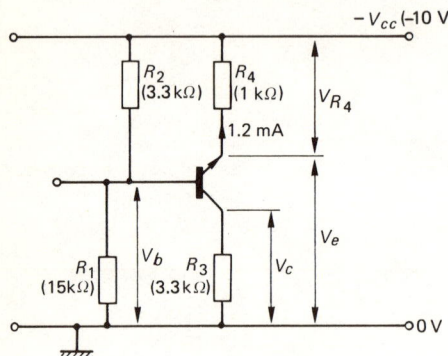

Fig.10.7 *npn transistor amplifier using negative supply* - V_{cc}

voltages are, therefore, negative as they are all measured with respect to the positive chassis. Using the typical values given in the diagram we get

$$V_b = \text{voltage across } R_1 = \frac{V_{cc}}{R_1 + R_2} \times R_1$$

$$= \frac{10}{15 + 3.3} \times 15 = \frac{10}{18.3} \times 15 = 8.2 \text{ V}$$

$$V_c = \text{voltage across } R_3 = I_c \times R_3 = 1.2 \text{ mA} \times 3.3 \text{ k}\Omega$$

$$= 4 \text{ V (approx.)}$$

$$V_e = V_{cc} - V_{R_4} = V_{cc} - I_e \times R_4 = 10 - 1.2 \times 1$$

$$= 10 - 1.2 = 8.8 \text{ V.}$$

The static operating condition of the transistor is therefore: e -8.8 V, b -8.2 V, c -4 V.

Base Current

Base current I_b (Fig.10.8) flows from the emitter across
the b-e junction through R_1 to the positive d.c. rail.
Bias resistor R_1 thus has two currents flowing through it:
standing current I_s (which also flows through R_2) plus
the base current I_b (which does not flow through R_2).
The volt drop across R_1 thus increases by $I_b \times R_1$. Since
$V_{R_1} + V_{R_2} = V_{cc}$ then an increase in V_{R_1} would result in
an equal decrease in V_{R_2}, i.e. a drop in base voltage.
Under normal circumstances I_b is very small and may
usually be neglected. Thus V_b is assumed to be determined
by resistor chain R_1-R_2.

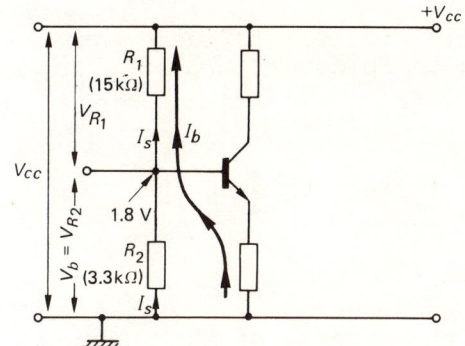

Fig.10.8 The effect of base current I_b

However, if the base current is large (e.g. when a transistor is conducting very heavily) or if R_1 is made very
large, the change in base voltage due to I_b begins to
affect the static conditions of the transistor and has to
be taken into account.

Consider the circuit in Fig.10.8. Under normal conditions a base current of say 10 µA produces a voltage
drop across R_1 of $I_b \times R_1 = 10 \times 10^{-6} \times 15 \times 10^3 =$
$150 \times 10^{-3} = 0.15$ V which is small compared with a base
voltage of 1.8 V determined by bias chain R_1-R_2. If the
transistor is now made to conduct very heavily, taking a
large current, base current will also be high say 80 µA.
Voltage across R_1 due to 80 µA base current is 80×10^{-6}
$\times 15 \times 10^3 = 80 \times 15 \times 10^{-3} = 1.2$ V. The base voltage
will drop by an equal amount i.e. from 1.8 V to 0.6 V.

Base Current Biasing

The base current can be used to provide the normal bias
for a transistor as shown in Fig.10.9. In this circuit
R_2 is dispensed with and a very large R_1 is used. Current
I_b is now entirely responsible for the volt drop across
R_1 (there is no standing current). This volt drop is
large enough to provide the normal bias.

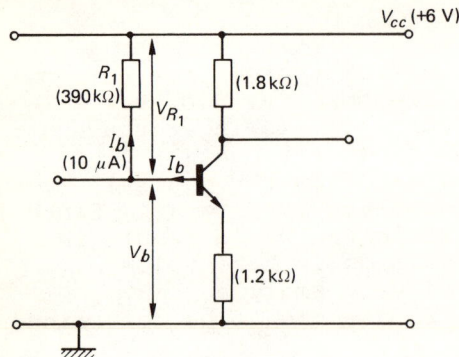

Fig.10.9 Base current biasing

Given that the base current in Fig.10.9 is 10 µA the base voltage may be calculated as follows:

$$V_{R_1} = I_b \times R_1 = 10 \times 10^{-6} \times 390 \times 10^3 = 3900 \times 10^{-3}$$
$$= 3.9 \text{ V}$$

The voltage at the base is the voltage between the base and chassis, i.e.

$$V_b = V_{cc} - V_{R_1} = 6 - 3.9 = 2.1 \text{ V}$$

The circuit in Fig.10.9 has the advantage of having a higher input impedance due to the absence of the input shunting resistor R_2.

Cut-off and Saturation

When the transistor is not functioning normally, i.e. when its static operating conditions are other than those specified by the design, it may be in one of two states, namely cut-off or saturation. These limiting or extreme states of a transistor may be caused by a faulty component in the circuit.

A transistor is said to be at cut-off when it ceases to conduct, i.e. its current is zero. With $I_e = 0$, no voltage is developed across R_4 (Fig.10.10). Hence the emitter voltage ($V_e = I_e \times R_4$) is also zero. A zero collector current produces no p.d. across R_3 ($V_{R_3} = I_c \times R_3$) making the collector go to the supply V_{cc}. Voltage between collector and emitter V_{CE} (= $V_c - V_e$) is, therefore, equal to V_{cc}.

A transistor is said to be saturated when the current through it is so high that it cannot increase any further, i.e. when I_e and I_c are at their maximum value. As I_e increases, V_e also increases (Fig.10.11). The voltage across R_3 increases as I_c increases, forcing the collector to go away from V_{cc} towards the emitter ($V_c = V_{cc} - V_{R_3}$). Hence as the current through the transistor increases,

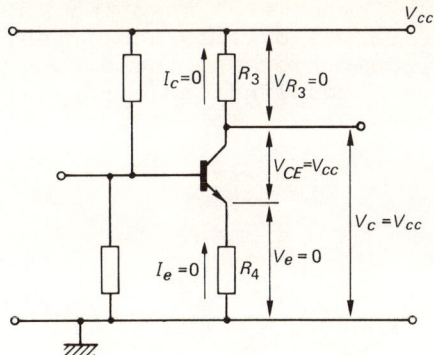

Fig.10.10 *Cut-off condition:*
$$V_e = 0 \text{ V}, \quad V_c = V_{cc}$$

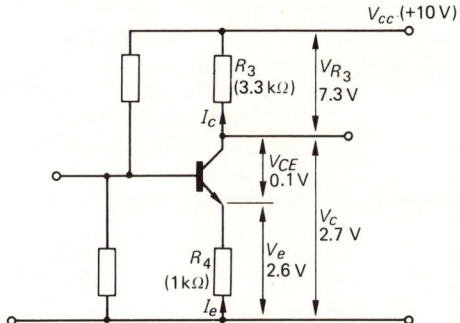

Fig.10.11 *Saturation condition:*
$$V_e \simeq V_c$$

the emitter voltage goes towards the collector while the collector goes towards the emitter. At saturation, when the transistor current is at a maximum, V_e and V_c are approximately equal, with V_{CE} almost at zero. Fig.10.11 gives typical values for saturation.

11 Common Emitter Amplifier - A.C. Operation

Coupling Capacitor

The purpose of a.c. coupling is to block the d.c. while allowing the a.c. signal through. Such coupling devices, e.g. a capacitor or a transformer, are usually used at the input and output of an amplifier. The quiescent or d.c. conditions of the transistor are thus not affected by the preceding or succeeding stages.

Fig.11.1 shows capacitor C coupling point A to point B with R as the load resistor. As far as d.c. is concerned, the capacitor acts as an open circuit blocking the d.c. between A and B. For this reason a coupling capacitor is also known as a *blocking capacitor*.

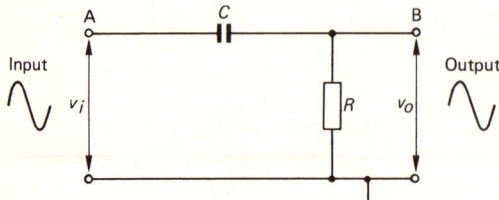

Fig.11.1 *Capacitor coupling*

For satisfactory a.c. coupling, the capacitor must have a reactance X_c at the operating frequency which is very small as compared with the value of the load resistor R so that a very small fraction of the input signal is dropped across C. For instance if V_1 = 100 mV, then satisfactory coupling may be said to be achieved if the output V_O = 95 mV, a drop of 5 mV (5%) due to the coupling capacitor.

The value of a coupling capacitor depends on two factors.

(1) LOAD RESISTOR R. Assuming that satisfactory coupling is achieved when $X_c = R/20$ and given that R = 1 kΩ and the operating frequency f = 300 Hz then,

$$X_c = \frac{1}{20}R = \frac{1}{20} \times 1000 = 50\Omega$$

From $X_c = 1/2\pi f C_1$, we get

$$C_1 = \frac{1}{2\pi f X_c} = \frac{1}{2\pi \times 300 \times 50} = 10.6 \ \mu F \ \text{or} \ 10 \ \mu F \ \text{approx.}$$

If the load resistor is increased to 100 kΩ,

$$X_c = \frac{1}{20}R = \frac{1}{20} \times 100 \ k\Omega = 5 \ k\Omega$$

$$C_2 = \frac{1}{2\pi f X_c} = \frac{1}{2\pi \times 300 \times 5 \times 10^3} = 0.1 \ \mu F$$

Hence if the load resistor is increased by 100 times (from 1 kΩ to 100 kΩ), the coupling capacitor may be decreased by the same proportion (from 10 μF to 0.1 μF).

In general the higher the load resistor, the smaller the required value of the coupling capacitor.

(2) THE FREQUENCY OF OPERATION. Take the above example where satisfactory coupling was achieved when C = 10 μF and R = 1 kΩ with f = 300 Hz.

If the frequency is now increased to 300 kHz then with $X_c = R/20 = 50\Omega$

$$C_3 = \frac{1}{2\pi f X_c} = \frac{1}{2\pi \times 300 \times 10^3 \times 50} = 0.01 \text{ }\mu F$$

Hence if the frequency is increased by 1000 times (from 300 Hz to 300 kHz), the coupling capacitor may be decreased by the same proportion (from 10 µF to 0.1 µF).

In general for a given value of R a large coupling capacitor is required for low frequencies and vice versa.
For a range of frequencies the value of the coupling capacitor is determined by the lowest frequency. Considering the above example, a 10 µF capacitor calculated to provide adequate coupling at 300 Hz will provide more than adequate coupling at 300 kHz. On the other hand a 0.1 µF capacitor calculated to provide adequate coupling at 300 kHz does not provide anything like adequate coupling at 300 Hz.

Decoupling

Fig.11.2b shows a capacitor C decoupling a resistor R. Without the capacitor (Fig.11.2a) point A has a d.c. voltage of 10 V and an a.c. signal voltage of 10 mV. The capacitor being open circuit for d.c. does not therefore

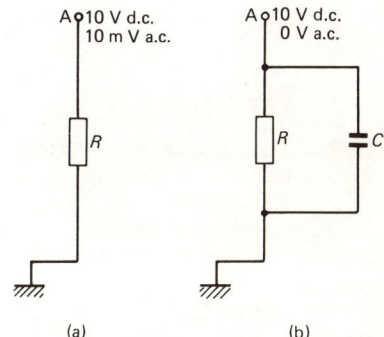

(a)   (b)

Fig.11.2 Effect of decoupling capacitor
        (a) Without decoupling capacitor
        (b) With decoupling capacitor

interfere with the d.c. voltage at point A. However, if the value of C is such that at the operating frequency its reactance is very small compared with R, all the a.c. signal will effectively be short-circuited to earth. Point A will thus have zero signal voltage. The value of C that would provide satisfactory decoupling is determined by resistor R as well as the operating frequency, in the same way as the value of the coupling capacitor described above.

R-C Coupled Amplifier

Fig.11.3 shows an $R-C$ coupled amplifier where C_1 is the
input coupling capacitor. The value of C_1 is compara-
tively large due to the low input impedance of the CE
transistor which is made even smaller by R_2 shunting the
input. C_2 couples the output to the load or the next
stage and is of a similar value to C_1. Typical values of
coupling capacitors are as follows:

for a.f., 10 µF - 50 µF; for r.f., 0.01 µF - 0.1 µF.

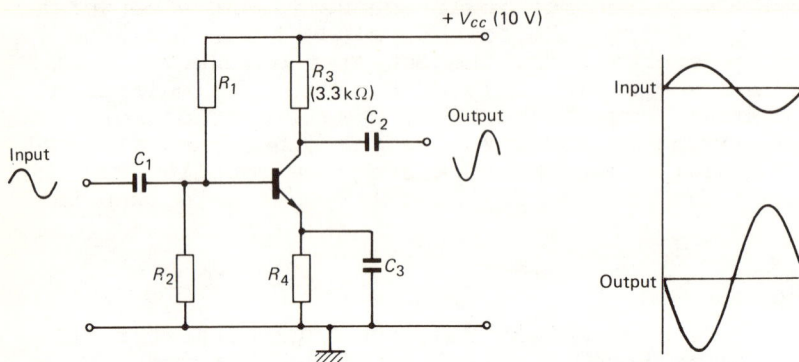

*Fig.11.3 R-C coupled amplifier with
decoupling capacitor C_3*

Fig.11.4 Phase reversal

Decoupling Capacitor

The negative feedback produced by R_4 in Fig.11.3 on the
one hand provides the necessary d.c. stability, while on
the other reduces the gain of the amplifier to a very low
level of the order of 2 to 3. The reduction in the gain
is due to the a.c. negative feedback provided by the
signal voltage drop across R_4. In order to remove the
a.c. negative feedback while retaining d.c. stability,
emitter decoupling capacitor C_3 is used.

Typical values for emitter decoupling capacitors are of
the same order as those for the coupling capacitor.

Amplification

The circuit in Fig.11.3 is the complete circuit for a
single stage CE amplifier. When an input signal, e.g. a
sine wave, is fed as shown C_1 couples the signal to pass
through to the base. For the positive half cycle, the
base goes away from the emitter, V_{be} goes up, I_e and
hence I_c increases, bringing V_c down. Thus the positive
half cycle of the input produces a negative-going half
cycle at the output. On the other hand the negative half
cycle of the input produces an upward or a positive movement in the collector voltage. The output is thus out of

phase with the input as shown in Fig.11.4. Amplification
occurs because a very small swing in V_{be} produces a large
swing in transistor current which, when passing through
R_3, produces a large voltage swing at the collector.

The Load Line

Transistor output characteristics show the general operation of the transistor. In order to represent the
operation of the transistor when used in a circuit, a
load line is drawn. Fig.11.5 shows the output characteristics for the transistor used in Fig.11.3 with XY as the
load line.

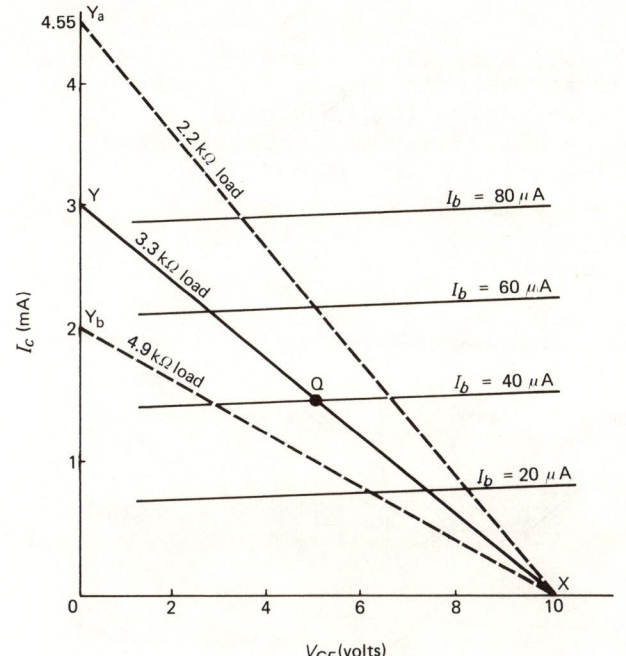

Fig.11.5 The load line

Before the load line can be drawn, two points falling
on it have to be located. The most convenient points are
point X on the x-axis where I_C = 0 and point Y on the y-
axis where V_C = 0. These two points are then joined
together to form the load line.

Point X. At this point, transistor current I_C is zero.
The transistor is at cut-off. Hence the collector
voltage V_C = V_{CC}.

Point Y. Here collector voltage V_C = 0. Putting V_C =
0 in the equation V_{CC} = V_C + V_{R_3}, we get V_{CC} = V_{R_3}. But
V_{R_3} = I_C × R_3, therefore V_{CC} = I_C × R_3, hence

$$I_C = V_{CC}/R_3$$

Using the values given in Fig.11.3, points X and Y can be
located as follows:

Point X. $I_c = 0$, $V_c = V_{cc} = 10$ V.

Point Y. $V_c = 0$, $I_c = V_{cc}/R_3 = 10/3.3 = 3$ mA

XY is therefore the load line for a load resistor R_3 of 3.3 kΩ.

Using a smaller load resistor, 2.2 kΩ, the load line is shifted to XY$_a$. Point X is the same as for the previous case since V_{cc} remains at 10 V. Point Y$_a$ however, is at $I_c = V_{cc}/R_3 = 10$ V/2.2 kΩ = 4.55 mA.

On the other hand a higher load resistor, e.g. 4.9 kΩ, would produce a load line XY$_b$ with point Y$_b$ at $I_c = 10$ V/4.9 kΩ = 2 mA.

Graphical Analysis

Amplification of a signal takes place along the load line and may be represented as shown in Fig.11.6. Point Q is the quiescent point representing the no-signal or d.c. operating point of the amplifier. At the Q-point the transistor is biased to have a base current

$I_b = 20$ mA with $I_c = 1.5$ mA and $V_c = 5$ V.

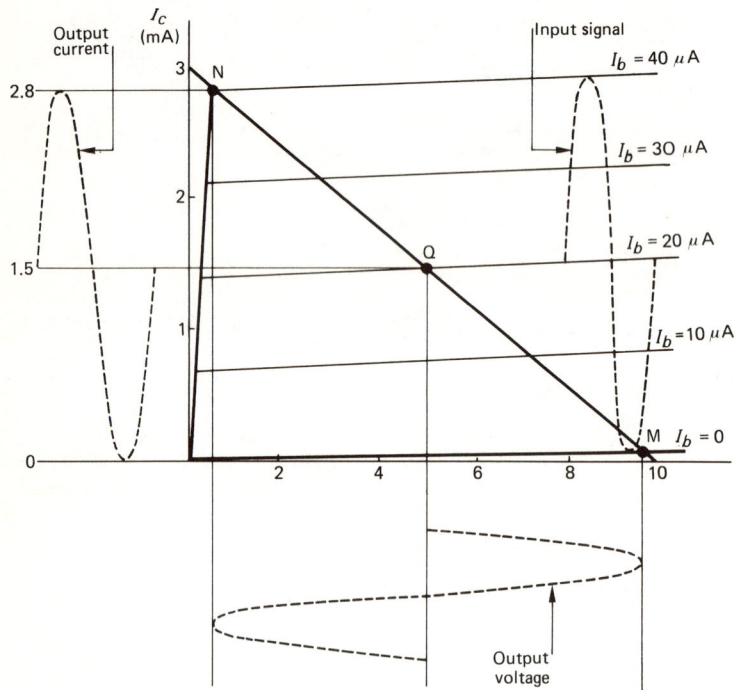

Fig.11.6 Graphical representation of amplifier operation

When a signal is applied, the base current swings as shown between 0 µA and 40 µA. This produces a collector output current I_c swing of 2.8 mA and a voltage swing of approximately 9 V.

The input swing is restricted on the one hand by the line I_b = 0 when the transistor is at cut-off (point M on the load line) and on the other by line I_b = 40 µA when the transistor is saturated (point N on the load line). In the case of the amplifier under consideration the Q-point is chosen in the middle so that by feeding a signal of a peak value of 20 µA into the base, the base current swings between zero and 40 µA producing the maximum undistorted output swing shown. Any attempt to enlarge the input signal will result in a distorted output waveform as shown in Fig.11.7, which is the case of an overdriven amplifier producing clipped sine waves. The input and output waveforms may be represented graphically using the transfer characteristics of Fig. 11.8. The operation is restricted to the linear part, otherwise distortion occurs.

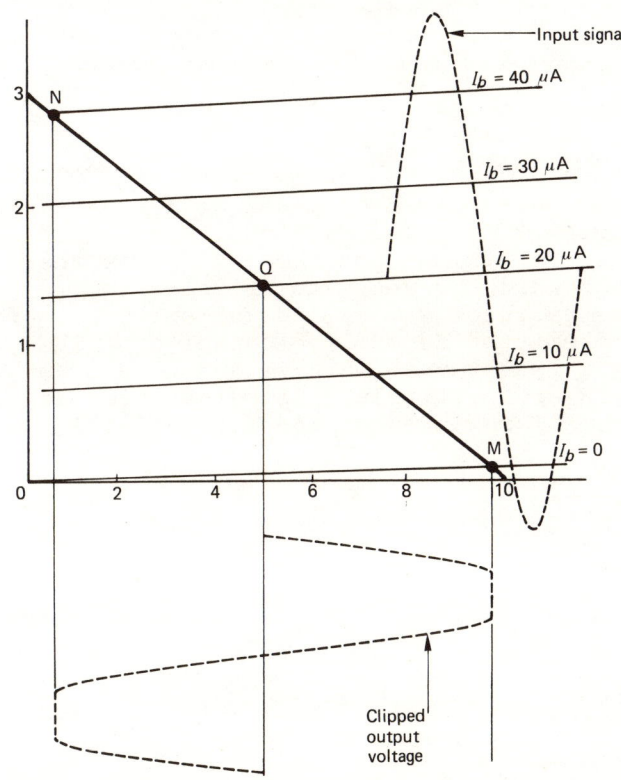

Fig.11.7 Overdriven amplifier producing clipped output

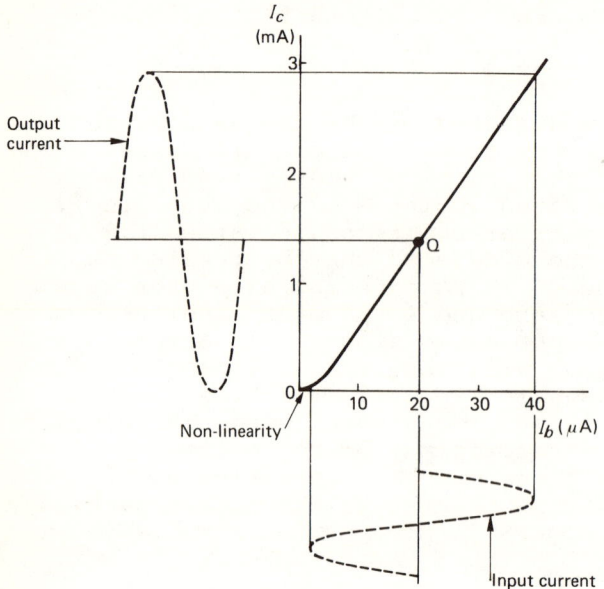

Fig.11.8 Graphical representation using the transfer characteristics

Transformer Coupled Amplifier

Interstage coupling can be achieved by means of a transformer as shown in Fig.11.9. R_1 and R_2 form a bias chain for TR_2 (the bias chain for TR_1 is not shown). C_2 is the bias coupling capacitor which prevents any signal from developing across bias resistor R_2. R_3 is the emitter resistor and C_2 is the emitter decoupling capacitor. The primary windings of the two transformers L_1 and L_3 act as the load for TR_1 and TR_2 respectively. Since transformer windings have a very small resistance, the d.c. voltage

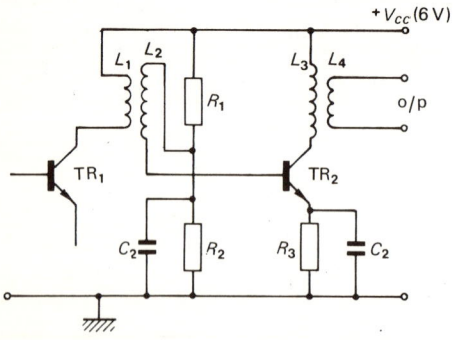

Fig.11.9 Transformer coupled amplifier

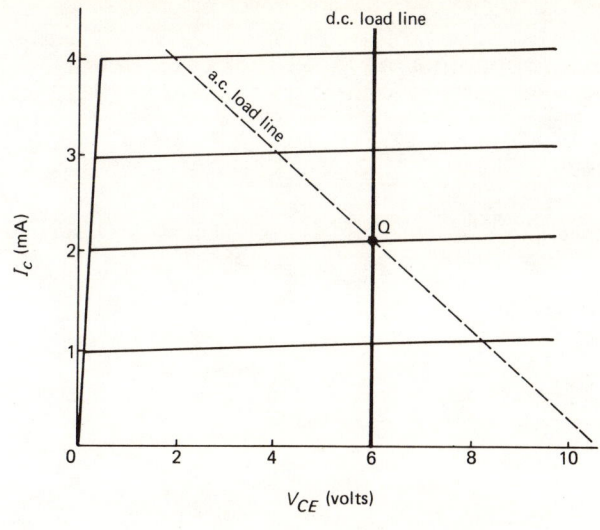

Fig.11.11
Transformer matching =
$r_1/r_2 = T_1^2/T_2^2 = n^2$

Fig.11.10 A.C. and d.c. load lines for
transformer coupled
amplifier in Fig.11.9

at the collector is at a constant value of V_{cc}. The
signal develops across the winding due to its a.c. resistance, i.e. its reactance. Two load lines can thus be
drawn as shown in Fig.11.10. If the supply voltage is
6 V, then the d.c. line is a vertical line representing
a constant voltage at the collector of 6 V. The a.c. line
represents the operation of the amplifier as the signal
is applied. The point of intersection is the quiescent
point Q.

COMPARISON BETWEEN R-C AND TRANSFORMER COUPLED AMPLIFIERS
$R-C$ coupling is used extensively in electronics due to
its high gain, better frequency response, cheap and small-
size components. Transformers, on the other hand, are
large in size and more expensive. Transformer coupling
has two main advantages:
(1) It does not use a collector resistance, with the
result that power dissipation, i.e. loss in the form of
heat, is small as compared with the $R-C$ coupled amplifier.
Transformer coupled amplifiers are, therefore, more
efficient.
(2) It facilitates matching between stages (see section
6). If r_1 is the output impedance of TR_1 and r_2 the
input impedance of TR_2 (Fig.11.11), then as was explained
in section 6

$$\frac{r_1}{r_2} = n^2 \quad \text{where } n \text{ is the turns ratio } \frac{T_1}{T_2}$$

Power Dissipation

Transistors like other components have a specific power
rating which must not be exceeded otherwise the transistor
might be damaged.

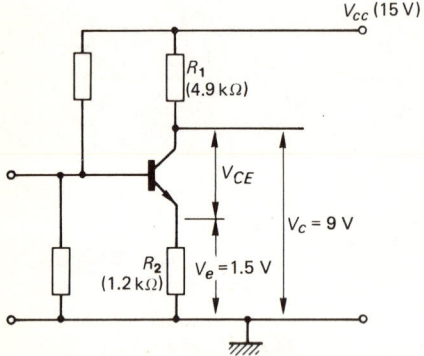

Fig.11.12 Power dissipation in a transistor = $I_c \times V_{CE}$

The power dissipated in a transistor is $I_c \times V_{CE}$ watts.
For example, referring to Fig.11.12, with $V_c = 9$ V,
$V_e = 1.5$ V,

$$V_{CE} = V_c - V_e = 9 - 1.5 = 7.5 \text{ V}$$

$$I_c = I_e = \frac{V_e}{R_2} = \frac{1.5 \text{ V}}{1.2 \text{ k}\Omega} = 1.25 \text{ mA}$$

Therefore power dissipation in the transistor is

$$I_c \times V_{CE} = 1.25 \text{ mA} \times 7.5 \text{ V} = 9.4 \text{ mW}$$

Note that power dissipated in load resistor is

$$I_c^2 \times R_1 = (1.25)^2 \times 4.9 = 7 \text{ mW}$$

The power rating of a transistor may be represented on
the output characteristic curve by a hyperbola or a
constant power dissipation curve as shown in Fig.11.13.
Points on this curve represent equal power dissipation.
The transistor must, therefore, operate on or below its
maximum rated power dissipation curve. To ensure maximum power output, the Q-point is usually chosen to fall
on the dissipation curve.

As can be seen from Fig.11.13, the operation of the
transistor is restricted by the saturation line, the
cut-off line ($I_b = 0$) and the rated power dissipation
curve. The area of operation is shown shaded.

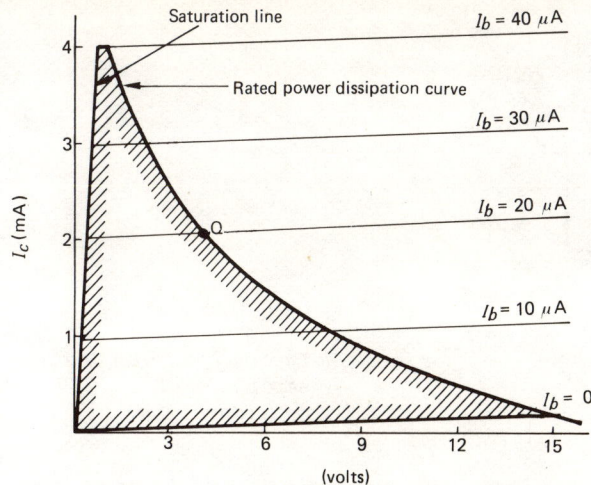

Fig.11.13 Operating area of a transistor (shown shaded)

12 The Common Base and Common Collector Amplifier

The CB Amplifier

Fig.12.1 shows an amplifier using a transistor in the common base configuration. Two separate e.m.f.s are used to provide the necessary bias. C_1 is a coupling capacitor feeding the input between the emitter and the base. The output is taken across the load resistor R_4.

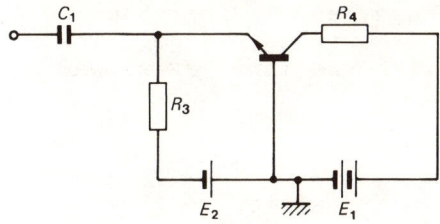

Fig.12.1 CB amplifier with E_1 and E_2 providing the bias voltage

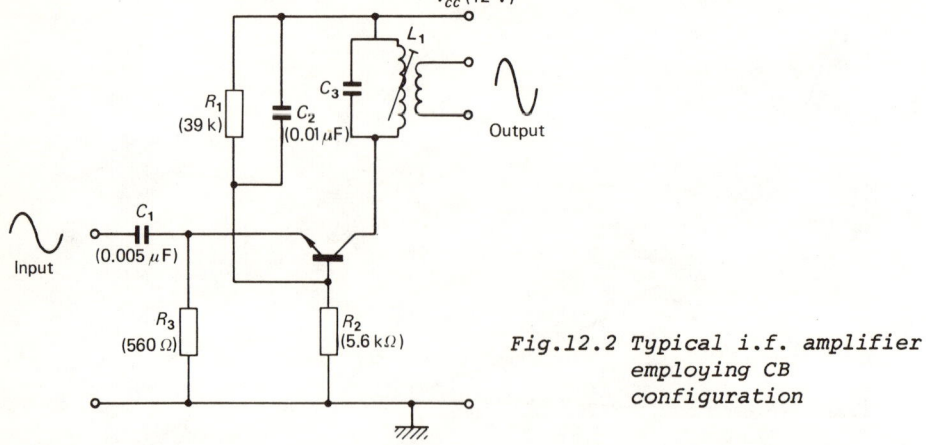

Fig.12.2 Typical i.f. amplifier employing CB configuration

Fig.12.2 shows a practical i.f. amplifier using one d.c. supply. The load is a tuned circuit C_3-L_1 employing transformer coupling, C_1 is the input coupling capacitor, R_1-R_2 forms the bias chain, and R_3 is the emitter resistor. C_2 is a decoupling capacitor ensuring that the base is at zero a.c. potential. In this case, the decoupling capacitor is taken to the supply line instead of the chassis. This is permissible since the supply line is at zero potential as far as a.c., i.e. the signal, is concerned. The difference between the supply line and the chassis is a d.c. potential only.

The CB amplifier has a low input impedance (50-100Ω) and a low gain compared with the CE configuration. It has the advantage, however, of a good high frequency response. Hence it is used at very high frequencies, e.g. r.f. amplifiers in radio and TV receivers, i.f. amplifiers for f.m. receivers, etc.

PHASE RELATIONSHIP If the input signal is negative-going, the emitter goes away from the base, and I_c increases to produce a drop in the collector or output voltage. The output is, therefore, in phase with the input as shown in Fig.12.2.

The CC Amplifier

Fig.12.3a shows a circuit using a transistor in the common collector configuration. C_1 and C_2 are the input and output coupling capacitors and R_1-R_2 forms the bias chain for the base voltage. No collector resistor is used since the output is taken at the emitter. Fig.12.3b shows the way the circuit is drawn normally. The output develops across the emitter resistor R_3 and hence the circuit is known as an *emitter follower*. It has a high current gain with a voltage gain of just below one. The low voltage gain is due to the 100% negative feedback produced by R_3.

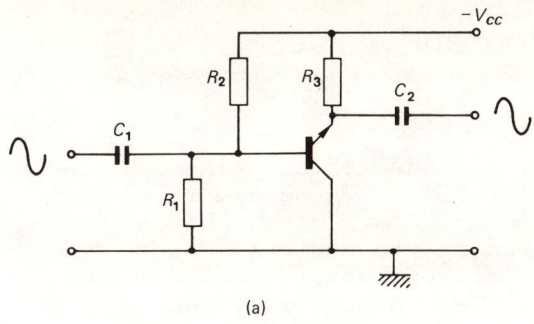

(a)

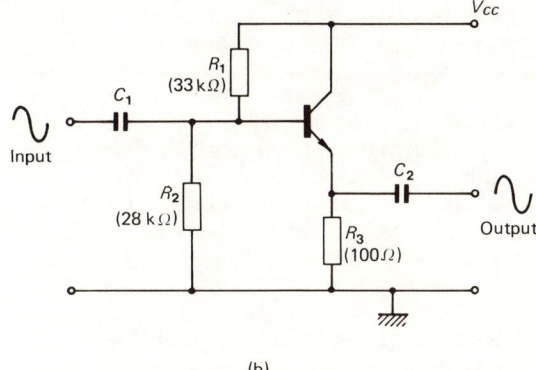

(b)

Fig.12.3 The emitter follower (or common collector) amplifier.
(b) shows the normal way in which the circuit is drawn

PHASE RELATIONSHIP When the input signal is positive-going, the base goes away from the emitter, i.e. V_b increases producing an increase in emitter or output voltage. Hence the output is in phase with the input.

COMPARISON BETWEEN CE, CB AND CC CONFIGURATIONS

Config-uration	Input Impedance	Output Impedance	Phase Reversal	Advantages
CE	1–2 kΩ	10–50 kΩ	Yes	High voltage and power gain
CB	Very low	Very high	No	Good high frequency response
CC	Very high	Very low	No	(*a*) Low output impedance (*b*) High current gain

13 Thermionic Emission and Valves

Electron Emission

As described in section 8, a conductor contains a large
number of free or loose electrons. These electrons may
leave the surface of the conductor if given enough energy
to overcome their attraction to the nucleus. Such energy
may be provided by heat, electron bombardment, or light.
The emitting conductor is always placed in vacuum to
facilitate the emission.
   There are three types of electron emission.
(1) *Thermionic emission* which occurs when a conductor is
heated. If sufficient energy is given to the electrons,
they leave the surface. This type is sometimes known as
primary emission.
(2) *Secondary emission* which occurs when high-velocity
electrons strike a conductor in a vacuum, knocking out
one or more electrons from the surface.
(3) *Photo-emission* occurring when energy in the form of
light falls upon the surface, knocking out electrons.

The Thermionic Diode

An electrode known as the cathode is placed inside a
sealed glass vacuum. The cathode is heated directly by
passing current through it, or indirectly (Fig.13.1) by a
heating filament. Electrons thus begin to be emitted
from the surface into the vicinity of the cathode, filling
the space around it with negatively charged electrons. A
space charge is then said to exist in the vicinity of the
cathode. As emission continues, the negative space charge
begins to repel the emitted electrons, and emission ceases.

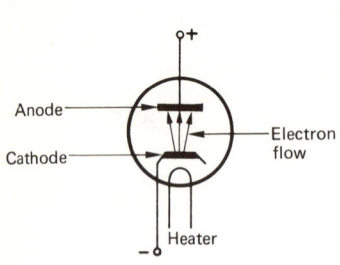

Fig.13.1 The thermionic diode

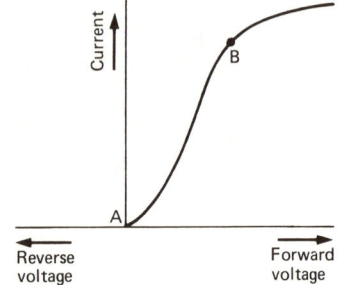

*Fig.13.2 Characteristics of
          a thermionic diode*

If a positive electrode known as the anode is placed
facing the cathode, electrons from the space charge will
be attracted to it and a flow of current from cathode to
anode will take place. The anode must be made positive
with respect to the cathode otherwise no current flows.
Hence the valve is a undirectional device known as a
thermionic diode.

The characteristics of the thermionic diode are shown
in Fig.13.2. From A to B the curve is similar to a pn
junction. At B the diode saturates when the anode
collects all the electrons emitted by the cathode. The
use of the thermionic diode is thus restricted to its
linear region AB. The symbol and the basic construction
of the thermionic diode are shown in Fig.13.3.

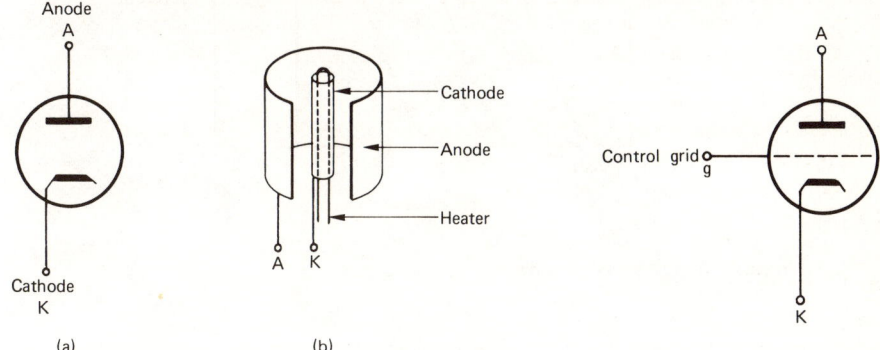

Fig.13.3 Thermionic diode   Fig.13.4 The triode
         (a) symbol, (b) construction

The Triode

A wire mesh known as the CONTROL GRID is placed between
the cathode and the anode, forming a triode (Fig.13.4).
The grid is placed in the space charge near the cathode.
The potential of the grid thus determines the space charge
density, which in turn controls the flow of electrons to
the anode. Because the grid is placed close to the
cathode, a small change in its potential produces a large
change in anode current, hence amplification.

Similar to the transistor, the triode may be used in
three different configurations: the Common Cathode, the
Common Grid, and the Common Anode (or the cathode follower). The Common Cathode is the most popular connection
and this will be dealt with in some detail.

Triode Characteristics

The transfer (or mutual) characteristics are shown in
Fig.13.5. They relate the output current I_a to the input
voltage V_g for specific values of anode voltage V_a. As
the grid goes negative it repels the electrons from the

64

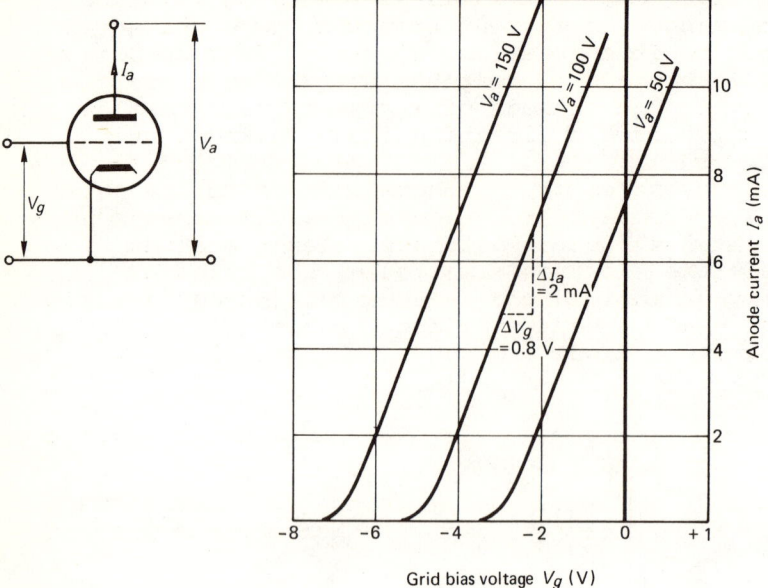

Fig.13.5 *Mutual characteristics of a triode*

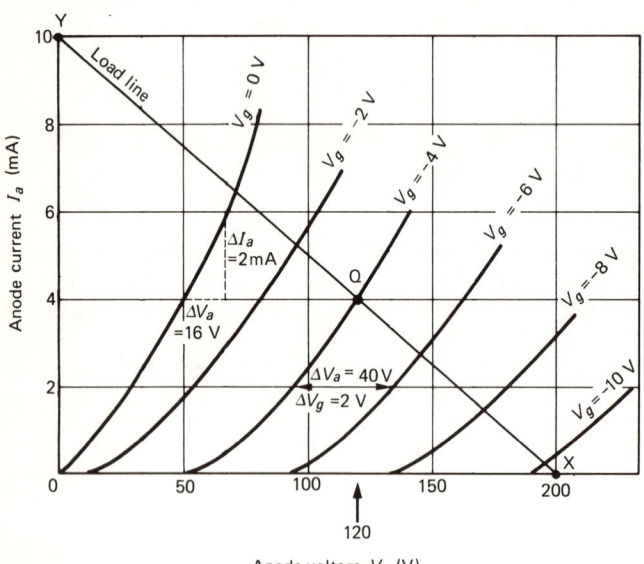

Fig.13.6 *Output characteristics of a triode*

cathode, reducing the anode current. Cut-off occurs when the grid negative potential is large enough to prevent the flow of any electrons to the anode. For small positive grid potential, electrons are speeded up towards the anode and I_a increases. For these small positive bias voltages, the transfer characteristics remain linear. However, higher positive bias will attract the electrons to the grid itself causing grid current to flow, a situation that must be avoided. For this reason the control grid is made negative with respect to the cathode.

The output characteristics are shown in Fig.13.6 relating the output current I_a to the output voltage V_a for specific values of grid bias.

Valve Parameters

There are three parameters that may be calculated for a valve from its characteristics.

$$\text{ANODE SLOPE RESISTANCE } r_a = \frac{\text{Change in anode voltage}}{\text{Change in anode current}}$$

$$= \frac{\Delta V_a}{\Delta I_a}$$

for a specific value of grid bias V_g.

Referring to Fig.13.6, r_a may be calculated by drawing a right angled triangle along the linear part of the characteristic at, say, zero grid bias.

$\Delta V_a = 16$ V and $\Delta I_a = 2$ mA giving $r_a = 16$ V/2 mA = 8 kΩ.

$$\text{MUTUAL CONDUCTANCE } g_m = \frac{\text{Change in anode current}}{\text{Change in grid voltage}}$$

$$= \frac{\Delta I_a}{\Delta V_g}$$

for a specific value of V_a.

Referring to Fig.13.5 draw a right angled triangle at, say, $V_a = 100$ V as shown.

$\Delta I_a = 2$ mA and $\Delta V_g = 0.8$ V giving $g_m = 2$ mA/0.8 V

$$= 2.5 \text{ mA/V}.$$

$$\text{AMPLIFICATION FACTOR } \mu = \frac{\text{Change in anode voltage}}{\text{Change in grid voltage}}$$

$$= \frac{\Delta V_a}{\Delta V_g}$$

for specific value of I_a.

Referring to Fig.13.6, for an anode current of 2 mA, $\Delta V_a = 40$ V and $\Delta V_g = 2$ V giving $\mu = 40$ V/2 V = 20 (no units).

From the above it can easily be deduced that

$$\mu = r_a \times g_m$$

In our example μ may be calculated from the values of r_a and g_m to give $\mu = 8$ k$\Omega \times 2.5$ mA/V $= 40$. When using this relationship, ensure that r_a is in kΩ and g_m in mA/V.

The Common Cathode Amplifier

Fig.13.7 shows an R-C coupled a.f. amplifier using a triode. The d.c. supply is usually referred to as the high tension HT voltage. Self-bias is used to give the grid the necessary negative potential with respect to the

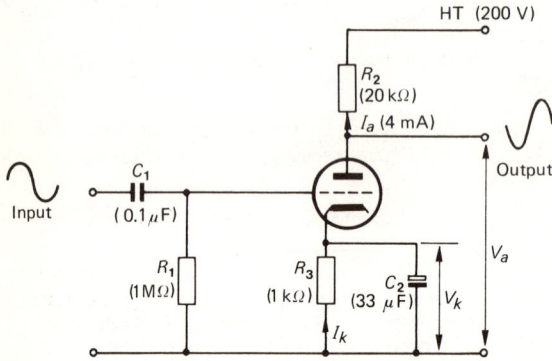

Fig.13.7 R-C coupled a.f. amplifier employing a triode

cathode. The cathode resistor R_3 is the resistor respon-sible for the self-bias. While the grid is kept at zero potential with respect to chassis, the cathode is made positive by the cathode current (I_k) passing through R_3. A positive cathode with respect to the grid is the same as a negative grid with respect to the cathode. Cathode resistor R_3 also functions as a d.c. stabiliser in the same way as an emitter resistor does in a transistor amplifier. C_2 is the cathode decoupling capacitor to remove the a.c. negative feedback. R_2 is the load resistor. R_1 is the grid leak resistor which provides a d.c. path for electrons that might settle at the grid. C_1 is a coupling capacitor.

One advantage of a valve over a transistor is that the former has a very high input impedance (in MΩ). Hence the value of the coupling capacitor C_1 is small, 0.1 μF. In the same way as a common emitter amplifier the output of the common cathode is in anti-phase to its input.

CALCULATIONS
Since the grid takes no current,

cathode current I_k = anode current I_a

Using the values given in Fig.13.7, we get cathode voltage

$V_k = I_k \times R_3 = 4$ mA $\times 1$ k$\Omega = 4$ V

With the grid at zero potential, the grid bias voltage is

$V_g - V_k = 0 - 4 = -4$ V

The voltage across the load resistor R_2 is

$I_a \times R_2 = 4$ mA $\times 20$ k$\Omega = 80$ V

But anode voltage $V_a =$ HT $-$ voltage across R_2.

Hence $V_a = 200$ V $- 80$ V $= 120$ V.

Referring to the output characteristics in Fig.13.6, a load line may be drawn as shown with Q as the quiescent point.

Point X on the load line is at $I_a = 0$ and $V_a =$ HT $= 200$ V.

Point Y is at $V_a = 0$ and $I_a = \dfrac{\text{HT}}{R_2} = \dfrac{200 \text{ V}}{20 \text{ k}\Omega} = 10$ mA

The Tetrode

The triode valve has a relatively low frequency response. This is due to the capacitance known as the inter-electrode capacitance between the three electrodes. To overcome this deficiency the tetrode was invented. A fourth electrode, in the form of wire mesh known as the SCREEN GRID, is placed between the control grid and the anode (Fig.13.8). It acts as an electrostatic screen (p.28) reducing the inter-electrode capacitance to very small values, improving the frequency response of the valve.

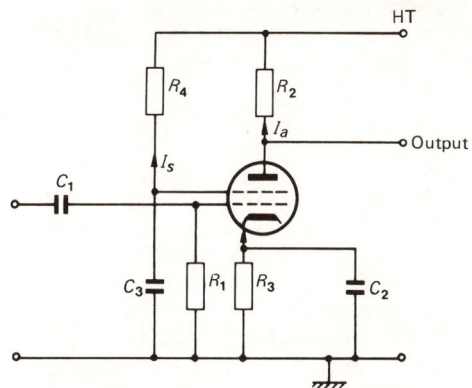

Fig.13.8 The tetrode amplifier

The screen is operated at a positive d.c. potential. Being nearer to the cathode it takes over the function of attracting the electrons from the cathode. However, due to the speed of the electrons reaching the screen, they pass through the wire mesh to be collected by the anode. The screen is normally decoupled to chassis as shown by C_3 to remove any a.c. signal (Fig.13.8) with R_4 as the screen potential resistor.

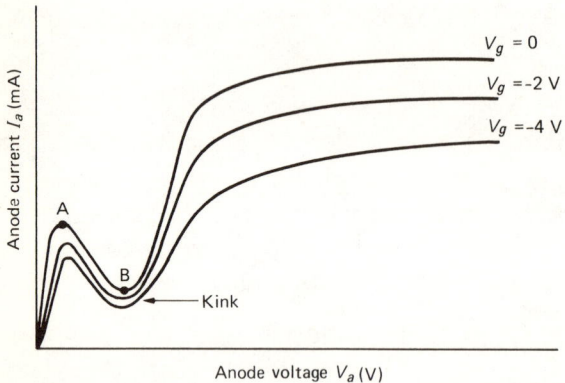

Fig.13.9 Output characteristics of a tetrode

Fig.13.9 shows the output characteristics of a tetrode. When the anode potential is lower than that of the screen, secondary emission of electrons from the anode are collected by the screen causing a drop in anode current and producing a kink in its characteristics. For the region between A and B the anode current decreases while its voltage increases. This is opposite to a normal resistor where its current increases as its voltage increases. Hence this region is known as a NEGATIVE RESISTANCE region. Devices such as the tetrode having negative resistance characteristics are capable of setting up oscillations without the aid of feedback. Tetrodes are usefully employed in oscillators.

The Pentode

As an amplifier the tetrode suffers from severe distortion due to the kink. To overcome this distortion, the pentode was constructed. A fifth electrode known as the SUPPRESSOR GRID is added between the screen and the anode (Fig.13.10). The suppressor is operated at chassis (or cathode) potential, thus preventing secondary emission electrons from reaching the screen. The output characteristics become linear as shown in Fig.13.11.

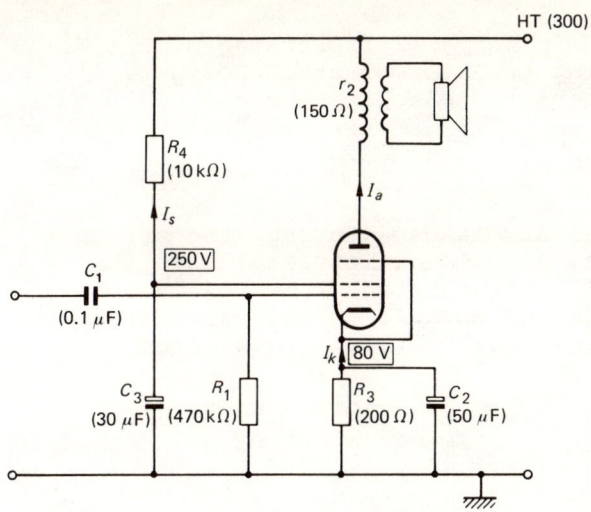

Fig.13.10 A.f. power amplifier employing a pentode

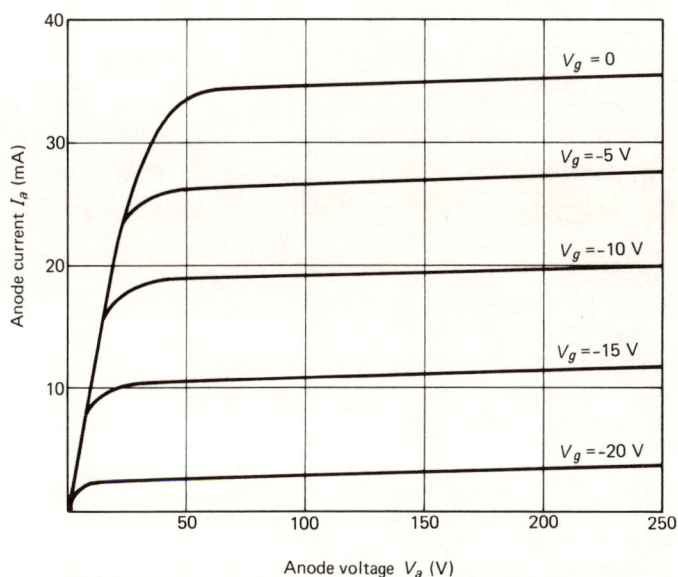

Fig.13.11 Output characteristics of a pentode

CALCULATIONS

Fig.13.10 shows an a.f. power amplifier using a pentode and employing a matching transformer at the output. Using the values given on the circuit we get

$$I_k = \frac{V_k}{R_3} = \frac{80 \text{ V}}{200\Omega} = 40 \text{ mA}$$

Although most of the electrons pass through the screen grid towards the anode, a small number settles on the screen, giving a small screen current I_s which may sometimes be neglected. In our example the screen current may be calculated as follows: voltage across screen resistor R_4 is

HT - screen voltage = 300 V - 250 V = 50 V.

Hence $I_s = \frac{50 \text{ V}}{R_4} = \frac{50 \text{ V}}{10 \text{ k}\Omega} = 5 \text{ mA}$

But $I_k = I_a + I_s$. Therefore

$$I_a = I_k - I_s = 40 - 5 = 35 \text{ mA}$$

The d.c. voltage across the primary winding of the transformer is

$$I_a \times r_a = 35 \text{ mA} \times 150\Omega = 52.5 \text{ V}$$

Hence the anode voltage is

300 V - 52.5 V = 247 V

The parameters of a pentode may be calculated in the same way as those for a triode. Compared with the triode, the pentode has a similar mutual conductance g_m but higher anode slope resistance r_a and higher amplification factor μ.

THE BEAM TETRODE The beam tetrode is another thermionic device which removes the non-linearity from the tetrode characteristics but without using a suppressor grid. Its characteristics and parameters are, therefore, very similar to those of the pentode.

14 The CRT and Other Thermionic Devices

The Cathode Ray Tube

The basis of the cathode ray tube was described in
Principles and Systems for Radio and TV Mechanics. In
this chapter further details are given. Fig.14.1 shows
the various electrodes in a CRT and their relative potentials. Electrons emitted from the cathode (or electron

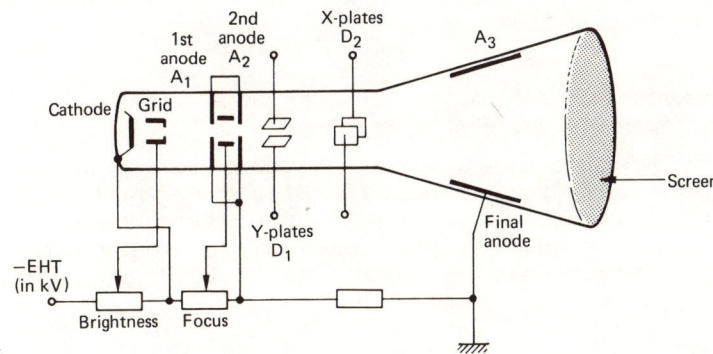

Fig.14.1 The cathode ray tube

gun) pass through a small hole (or aperture) in the grid.
The grid potential determines the intensity or number of
electrons emitted and thus the brightness of the trace on
the screen. The electron beam then passes through an
electron lens which focusses the beam on the screen. A_3
is the final anode with a potential of a few kV (with
respect to the cathode), referred to as the extra high
tension or EHT. Two pairs of deflecting plates D_1 and D_2
are used to provide electrostatic deflection of the beam
in the vertical Y and the horizontal X directions respectively.

Deflection Methods

Two methods are used to deflect the beam in a CRT. The
electrostatic method using two plates facing each other and
an electric potential across them (Fig.14.2a). An electrostatic field is thus created which deflects the electrons
passing through it. *Electromagnetic* deflection on the
other hand uses the magnetic field created by a current

passing through a coil to deflect the electrons. Two sets
of deflecting coils (known as scanning coils in a TV
receiver) are used as shown in Fig.14.2b. Both methods
are capable of producing linear deflection. However,

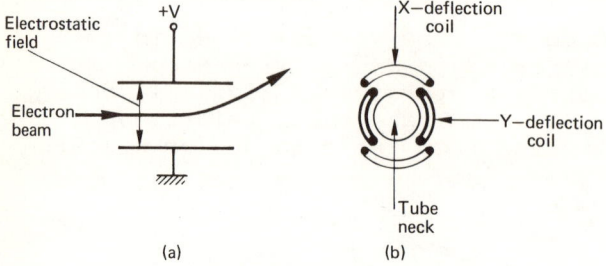

Fig.14.2 (a) Electrostatic deflection
        (b) Electromagnetic deflection

electrostatic deflection has a wide frequency range, hence
its use in cathode ray oscilloscopes. Electromagnetic
deflection is more suitable for high tube voltage such
as a TV receiver and requires less space since both coils
are placed at the same position along the neck of the tube.

The Gas-filled Diode

A small amount of gas under low pressure is added to an
ordinary thermionic diode to make the gas-filled diode
which is also known as a neon diode. A process known as
ionization takes place when the electrons emitted from
the cathode strike atoms of the gas inside the tube so
freeing its electrons. Ionization is accompanied by a
glow of the gas inside the tube. As soon as ionization
occurs the voltage across the diode drops to what is known
as the *holding voltage* (Fig.14.3) which the diode maintains constant. The voltage at which ionization occurs
is known as the *striking voltage* of the diode. Once fired,
the diode may be switched off only when the voltage across
it falls to its extinction voltage, i.e. when the anode
potential is almost down to that of the cathode. Gas-
filled diodes are used as shunt stabilisers (see section 19).

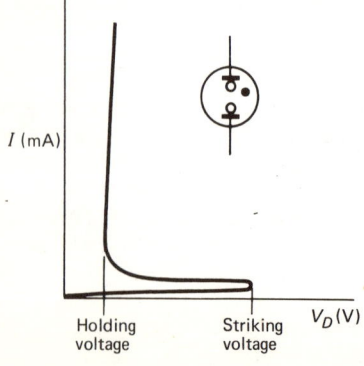

I (mA)

Holding
voltage

Striking
voltage

V_D(V)

Fig.14.3 Characteristic curve for a
         gas filled (neon) diode

The Thyratron

The striking voltage of a gas-filled diode may be varied
by introducing a control grid between the anode and
cathode as shown in Fig.14.4. The thyratron may thus be
considered as a gas-filled triode. The grid voltage

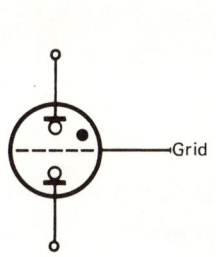

Grid

Fig.14.4 The thyratron

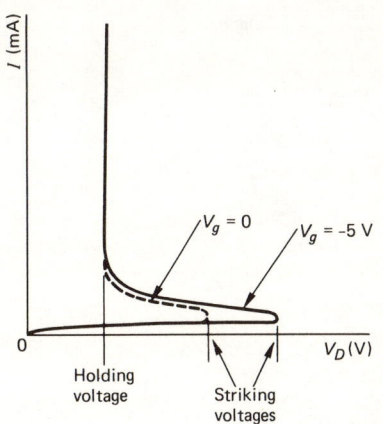

Fig.14.5 Characteristic curves
of a thyratron

determines the voltage at which breakdown or ionization
occurs (Fig.14.5). A negative grid delays ionization,
i.e. increases the striking voltage, while a positive
grid reduces the striking voltage. Once the thyratron
conducts, the grid has no effect on the anode current.
It can only be turned off by reducing the anode voltage
to the level of the cathode i.e. below the extinction
voltage.

U.H.F. and Microwave Valves

The time it takes an electron to travel from the cathode
to the anode, known as the transit time, limits the use
of valves at v.h.f., u.h.f. or higher frequencies. For
such high frequencies, i.e. 1000 MHz and above, special
vacuum tubes are used. The Klystron and the Magnetron
are two such valves.

   In the KLYSTRON a beam of electrons from the cathode is
made to pass through gaps in closed cavity resonators
built into the tube in order to make it possible to overcome the transit time limitations. The klystron used in
u.h.f. transmitters is capable of delivering an output
power of tens of kWA.

   The MAGNETRON is a vacuum tube whose current is influ-
enced by a magnetic field. The magnetic field is created
by a permanent magnet which together with the electric
field due to the electron beam makes it possible to deliver
powers in the region of 1000 kW in the u.h.f. range.

74

The Electron Ray Indicator

Also known as a magic eye or a tuning indicator, the
electron ray indicator is a cross between a triode and a
cathode ray tube. Fig.14.6 shows a schematic diagram for
this tube. The cathode is in the form of a cylinder with

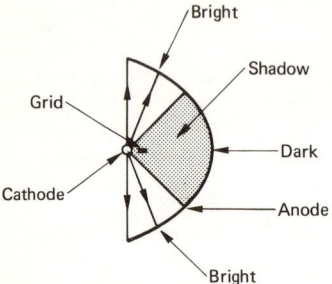

*Fig.14.6 Schematic diagram of an
electron ray indicator*

the grid as a metal strip and the anode in the shape of a
wide-angle cone coated with a fluorescent material. A
negative grid potential casts a "shadow" on the anode
allowing electrons to hit the two ends of the anode cone
and not the middle part. If the grid is made less negative, the shadow is narrowed and a larger area of the
anode is thus struck with electrons which begins to glow,
thus "closing the eye".

15 Photoelectric Devices

Photoelectric devices may be divided into four types.
(1) The PHOTO-EMISSIVE CELL or phototube, which employs
the principle of electron emission from a cathode when
light falls upon its surface.
(2) The PHOTOCONDUCTIVE CELL, which reduces its resistance as the intensity of light falling upon it is
increased.
(3) The PHOTOVOLTAIC CELL which generates an e.m.f.
according to the intensity of light falling upon its
surface. Semiconductor material such as selenium may be
used; it produces a high enough voltage to deflect a
voltmeter and can be used in lightmeters. The photovoltaic cells are also used to provide power for telephone
repeaters and for communication satellites. They form
the basis of solar generators where sunlight is used to
generate electric power.

(4) PHOTODIODES and PHOTOTRANSISTORS. In the photodiode
a pn junction is reverse biased. The only current passing
through the junction is, therefore, a small leakage
current, which is a nuisance when the junction is used
normally. In the photodiode it becomes the working
current. It was found that the leakage current not only
increases with heat but also with light waves falling
upon the junction. That is why diodes and transistors
are packed inside opaque containers. Photodiodes on the
other hand are provided with a small window through which
light can fall on the junction increasing its leakage
current.

Phototransistors use the same principle as the photodiode. The junction in this case is the reverse biased
base-collector junction. The advantage of the photo-
transistor is that the increased b-c leakage current due
to light is amplified by the transistor which may then
provide enough power to deflect a voltmeter as in the
case of a lightmeter. Fig.15.1 shows a simple phototrans-
istor circuit. Multiple phototransistors may be used to
drive a motor, etc.

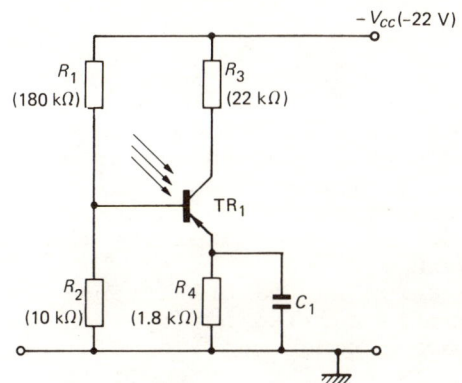

Fig.15.1 Circuit using a
photo-transistor

THE PHOTOMULTIPLIER. Electron multiplication can be
achieved by secondary emission. One such device is the
photomultiplier, shown in Fig.15.2. A photo-emissive
cell is used to emit a small number of electrons from
the cathode. These electrons are used to bombard a
treated surface of a positive electrode E_1, known as the
dynode. Secondary emission takes place with a larger
number of electrons now being emitted from E_1. This
process is then repeated a number of times, and each
time the number of electrons emitted increases by a
factor determined among other things by the coating of
the dynodes. A final positive anode A is used to collect
the electrons, forming an anode current I_a which is fed
into the load.

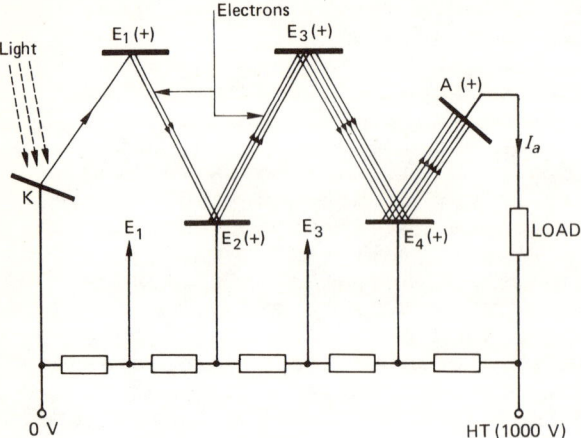

Fig.15.2 Photomultiplier

16 Field Effect Transistors

The field effect transistor (fet) is a semiconductor
device that is becoming very popular in modern electronic
equipments. It consists basically of a semiconductor
current-carrying channel, the resistance of which is
controlled by an electric field, thereby controlling the
flow of current through the channel.

Field effect transistors are known as UNIPOLAR transistors since the flow of current is carried out by the
majority carriers only. It flows through one type of
semiconductor, an n-type or a p-type. The action of the
normal junction transistor on the other hand is based on
minority as well as majority carriers. This is because
its current flows through a forward biased b-e junction
(majority carriers) and a reverse biased b-c junction
(minority carriers). Hence they are known as BIPOLAR
transistors.

Field effect transistors have three electrodes known as
the source (s), the gate (g), and the drain (d). They
correspond to the emitter, base and collector for the
bipolar transistor, and the cathode, control grid and
anode for a valve.

Field effect transistors combine most of the good
points of valves and the bipolar transistors. They require
no heating and are small in size. They have very high
input impedance, as high as that of a valve. They are
less temperature-sensitive than the bipolar transistors
and hence less prone to thermal runaway. Circuits using

fets are simple in design, using less components than an equivalent circuit using bipolar transistors.

Field effect transistors are simple to manufacture, and more amenable for use in integrated circuits than their counterparts, the bipolar transistors.

There are two main types: the junction field effect transistor and the MOS field effect transistor.

The Junction FET

Consider an n-type semiconductor channel with a d.c. potential V_{DD} connected across it as shown in Fig.16.1a. A current known as drain current I_D flows through the

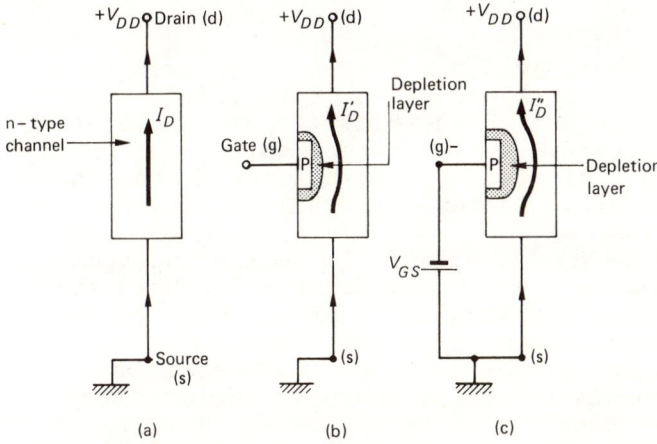

Fig.16.1 *The operation of a junction fet*

channel from source to drain as shown. If now a p-type semiconductor known as the gate is diffused into the n-type channel as shown in 16.1b, a pn junction is formed. At the junction a depletion layer is established in the same manner as a normal pn junction. As can be seen, the depletion layer restricts the flow of current through the channel by reducing the effective width of the channel. In other words it increases the resistance of the channel. The depletion layer may be widened, thus restricting further the flow of current by applying a reverse bias V_{GS} to the junction as shown in 16.1c. By changing the bias voltage at the gate, drain current I_D may be varied. Fig.16.2 shows a cross-section of the structure of a field effect transistor.

FETs with p-type channel may also be used, with negative d.c. supply $-V_{DD}$. The symbols for both types of junction fets are shown in Fig.16.3.

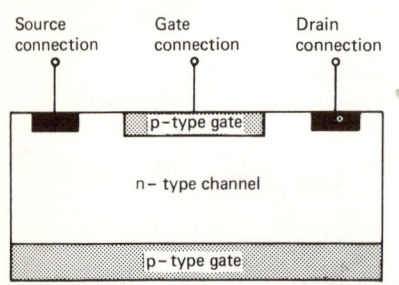

Fig.16.2 Cross-section of the structure of a junction fet

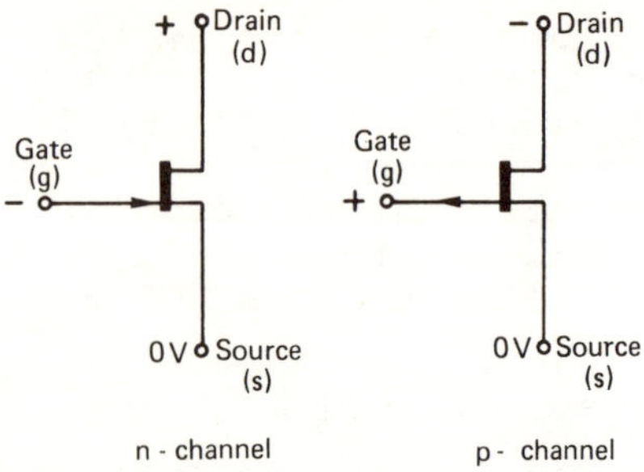

Fig.16.3 Junction fet symbols

Output Characteristics

The output characteristics for a junction fet used in the common source configuration is shown in Fig.16.4. They are similar to the output characteristics of a bipolar transistor or a pentode. They show how output current I_D changes with output voltage (voltage between drain and source) V_{DS} for specific values of gate bias (voltage between gate and source) V_{GS}.

The change in gate-source bias is large (in the region of a few volts) unlike the bipolar transistor where the base-emitter voltage is basically constant. The fet is therefore similar to a valve in this respect.

It can be seen that as the gate bias increases, so drain current decreases, until the depletion layer due to the gate-source junction fills up the whole width of the channel, stopping the flow of current. The fet is then said to be cut off.

Pinch-off Voltage

Consider the output curve with V_{GS} = O (Fig.16.4). As the drain voltage I_D is increased from zero, drain current increases gradually until point P, after which it stays constant. The voltage at point P is known as the pinch-off voltage. At this point the depletion layer due to the reverse biased gate-drain junction has extended right across the channel. Drain current I_D does not stop flowing, however, because the depletion layer is created precisely because of the drain current itself. For grid bias of -1 V, -2 V, and so on, similar pinch-off points are observed P_1, P_2, etc. Joining these points together forms the pinch-off region where the fet is made to operate.

The FET Common Source Amplifier

A typical a.f. amplifier using a field effect transistor is shown in Fig.16.5. As can be seen, it is very similar

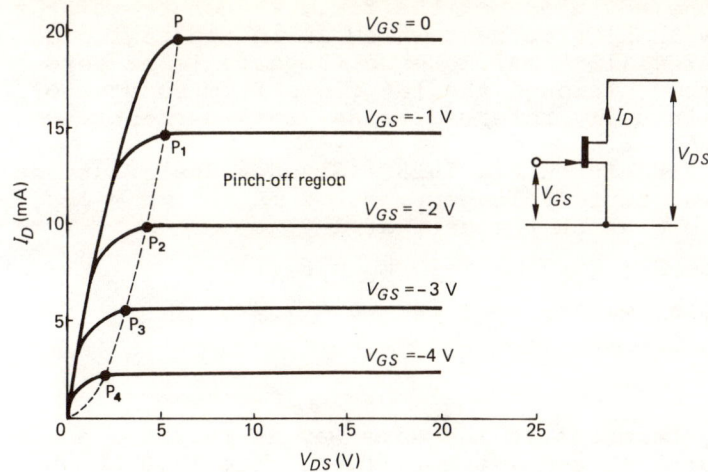

Fig.16.4 *The output characteristics of a junction fet*

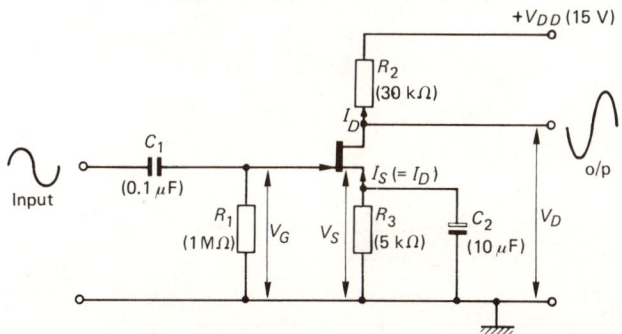

Fig.16.5 *A.f. amplifier using an n-channel junction fet*

to a valve amplifier. R_1 is a leakage resistor; R_3 provides the necessary reverse bias by giving the source a
positive voltage above that of the gate. R_3 also provides
d.c. stability for the amplifier. R_2 is the load resistor
which may be as high as 1.5 MΩ. C_2 is the source decoupling capacitor removing the a.c. negative feedback caused
by R_3. It will be noticed that a low value coupling
capacitor 0.1 µF is used. This provides adequate coupling
due to the high input impedance of the fet.

When a signal is applied into the input, the fet drain
current will vary accordingly, producing an output voltage
at the drain. For the positive half cycle, the gate goes
in the positive direction, thereby reducing the reverse
bias of the gate-source junction, and thus increasing the
fet current I_D. As I_D increases, drain output voltage
decreases and a negative-going half cycle is produced.
On the other hand, the negative half cycle of the input
produces a positive-going half cycle at the output. Hence
the output is in anti-phase to the input.

CALCULATIONS   One advantage of the field effect transistor is its very low leakage current which is in the region of a few pico-amperes (10^{-6} mA).   Hence the gate is at zero potential.   Current through the fet flows from source to drain and it is usually referred to as drain current I_D (= source current I_S).

Consider the circuit in Fig.16.5.   Given a drain current I_D = 0.2 mA then source voltage V_S = 0.2 mA × 5 kΩ = 1 V, giving a reverse bias at the gate of 1 V.

Voltage across R_2 = 0.2 mA × 30 kΩ = 6 V.

Therefore drain voltage V_D = 15 - 6 = 9 V.

The Load Line

A load line may be drawn in the same way as that for a bipolar transistor or a valve amplifier.   Fig.16.6 shows the load line for the amplifier in Fig.16.5.

When I_D = 0, V_{DS} = V_{DD} = 15 V giving us point X on the load line.

When V_{DS} = 0, the total d.c. supply V_{DD} falls across R_2.

Hence $I_D = \dfrac{V_{DD}}{R_2} = \dfrac{15 \text{ V}}{30 \text{ k}\Omega}$ = 0.5 mA giving us point Y.

The quiescent point Q is chosen so that the transistor operates within its pinch-off region.

Quiescent or operating point Q is hence at I_D = 0.2 mA and V_{GS} = -1 V with V_{DS} = 9 V.

MOSFET

This type of field effect transistor has a metal gate which is electrically insulated from the semiconductor, by a thin oxide film.   Hence its name MOS, which stands for Metal Oxide Silicon.

The n-type channel is formed by the gate insulating oxide attracting electrons from a p-type substrate (Fig. 16.7).   The thickness of the n-channel may be varied by applying a voltage to the gate.   A positive voltage widens the n-type channel increasing the current, while a negative voltage narrows the n-channel and reduces the current. The opposite is true for a p-channel type.

There are two types of mosfets: the enhancement (normally off) and the depletion (normally on).   In the ENHANCEMENT type the fet is at cut-off (normally off), when gate bias V_{GS} = 0.   Current flows only when a bias voltage is applied to the gate.   The output characteristics for an n-channel enhancement type are shown in Fig.16.8, together with its symbol.

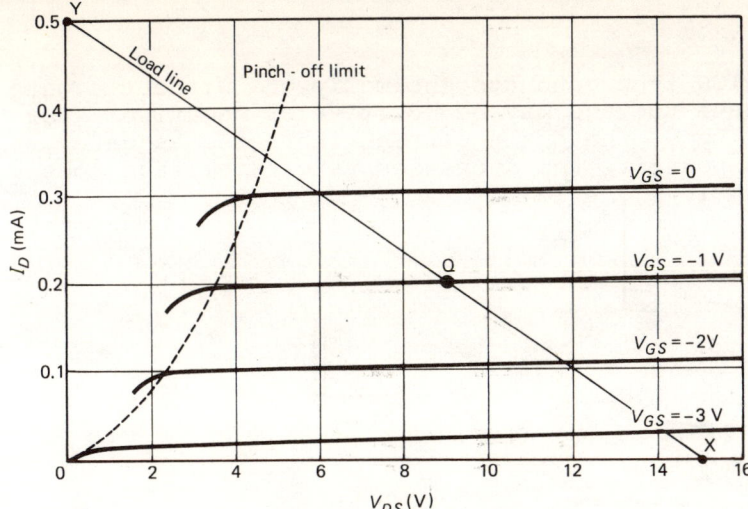

Fig.16.6 Load line for the fet amplifier in Fig.16.5

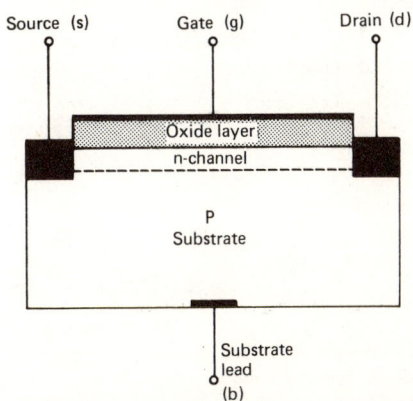

Fig.16.7 Cross-section of a mosfet

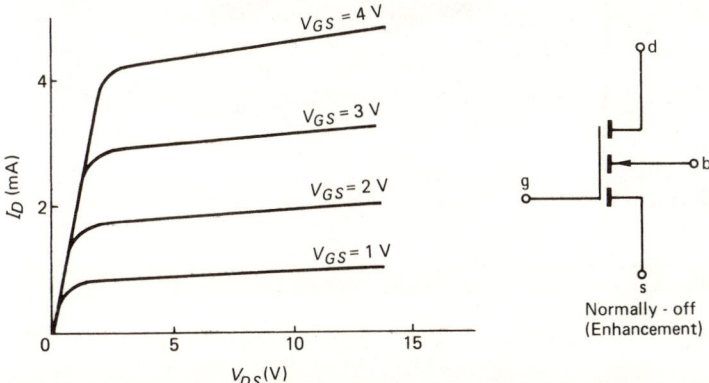

Fig.16.8 The output characteristics of a normally-off (enhancement)
type n-channel mosfet. (Symbol is also shown)

The DEPLETION type conducts (normally on) without a gate bias. The gate voltage may be positive or negative. For an n-channel type, the drain current increases as gate voltage goes positive and decreases as gate voltage goes negative (Fig.16.9).

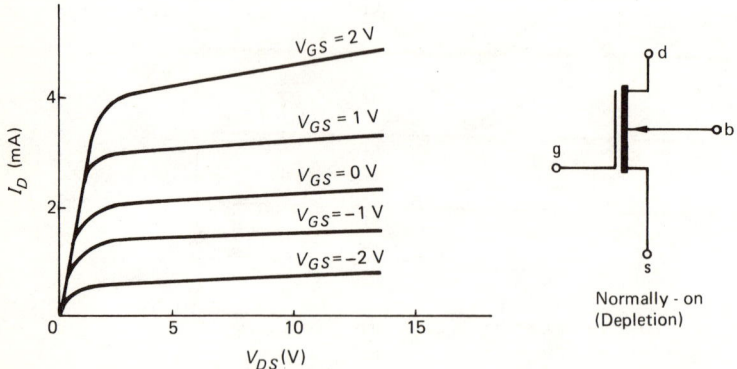

Fig.16.9 *Output characteristics of a normally-on (depletion) type n-channel mosfet. (Symbol is also shown)*

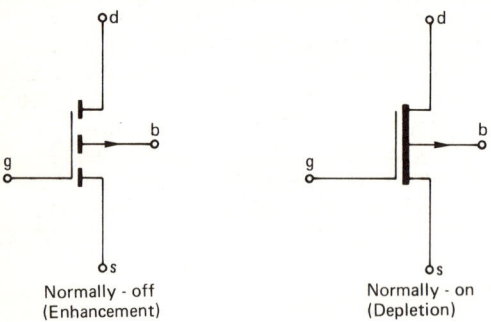

Fig.16.10 *Symbol of p-channel mosfet*

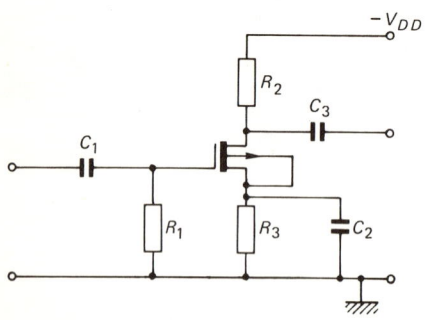

Fig.16.11 *FET amplifier using a p-channel depletion type*

The symbols for p-channel mosfets are shown in Fig.16.10. Note that the broken lines indicate a normally off (i.e. enhancement) type. A continuous line, on the other hand, indicates a normally on (depletion) type. Contact b is the substrate lead which is normally taken to source potential. Fig.16.11 shows a typical amplifier using a p-channel depletion mosfet in the common source configuration. A negative d.c. supply is used. The positive bias between the gate and source V_{GS} is produced in the normal way by source resistor R_3.

17 More Solid State Devices

The Tunnel Diode

The tunnel diode is a pn junction using heavily doped semiconductors and a very narrow depletion layer. Because of the heavy doping, the tunnel diode conducts in the reverse as well as in the forward direction. Hence it is only used when forward biased.

Fig.17.1 shows the forward characteristics for a tunnel diode. As the forward bias is increased from zero, the diode current increases sharply up to a maximum at point P. Further increase in the voltage causes the current to decrease to a minimum value (point M) after which the current begins to rise again. For the region between P and M the diode has a negative resistance (similar to the tetrode), which is made use of in high frequency oscillators and switching devices.

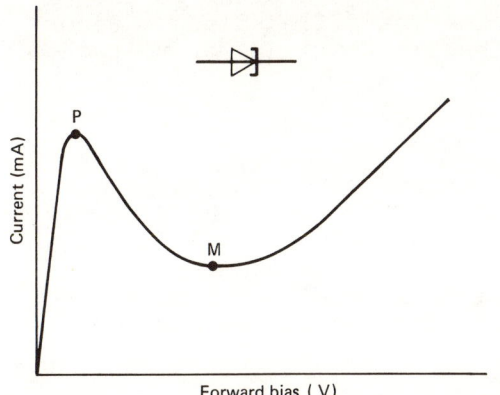

Fig.17.1 Symbol and forward characteristics of a tunnel diode

The Zener Diode

The zener diode is a junction diode which has a specified reverse breakdown voltage. Unlike the ordinary diode the zener is operated at its breakdown voltage. Fig.17.2 shows the characteristics for a zener diode. In the forward direction it behaves in the same way as a normal

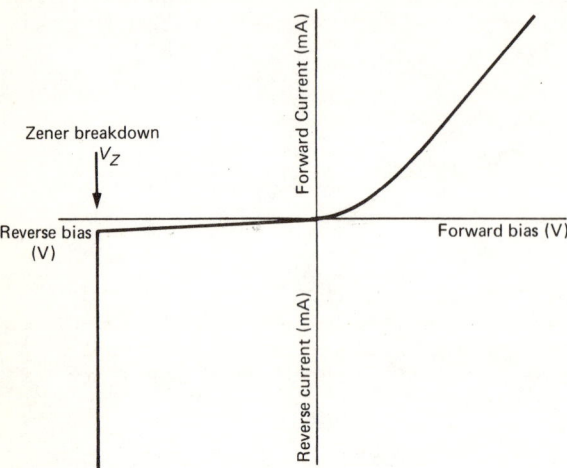

Fig.17.2 Forward and reverse characteristics of a Zener diode

diode. When reverse biased, no current flows until the zener reaches its breakdown voltage V_Z, usually known as the zener voltage, when current suddenly begins to flow. At breakdown the voltage across the zener remains constant for a very large change of current. The zener is the semiconductor equivalent of the gas-filled diode. Zener diodes are used for shunt stabilisation and for voltage reference (see section 19).

The Switching Diode

The switching diode consists of four layers of semiconductor material as shown in Fig.17.3. When the diode is forward biased a very small current flows until breakdown occurs (Fig.17.4). Before breakdown the diode may be considered as a switch in the "off" position. After breakdown it may be considered as a switch in the "on" position.

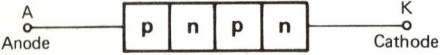

Fig.17.3 Switching diode

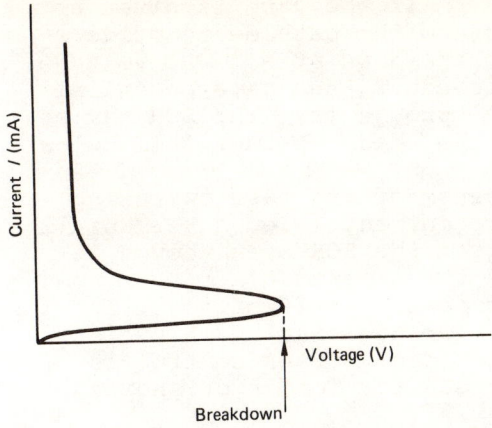

Fig.17.4 Characteristics of a switching diode

The Controlled Rectifier

The controlled rectifier or the THYRISTOR is another
4-layer pnpn junction device usually made of silicon;
hence the more common name, the silicon controlled rectifier SCR.   Unlike the switching diode the SCR has a third
terminal known as the gate (Fig.17.5).   The critical
breakdown voltage may now be varied by the potential at
the gate.   Fig.17.6 shows the characteristics of an SCR
for two different values of gate current.   At zero gate
current (i.e. a gate potential of zero volts), the SCR

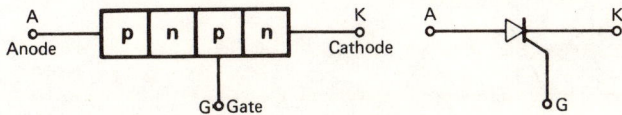

Fig.17.5 The SCR and its symbol

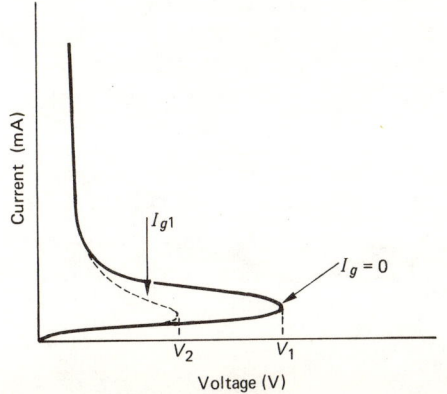

Fig.17.6 Characteristics of the SCR

has a breakdown voltage of V_1. If the gate is given a positive voltage with respect to the cathode to drive a gate current I_{g1}, the SCR breaks down at a lower voltage V_2. Once the SCR starts to conduct, the gate potential has no effect on the current through it. The SCR may be switched off only by making the anode voltage fall below the cathode potential.

The SCR is widely used because of its fast switching and because a very small gate current (i.e. a very small power) is sufficient to trigger the SCR even though it may be handling a current of few amperes.

It is extensively used for power rectification. It turns on during the positive (or negative) half cycle only, thus producing pulsating d.c. Power control is achieved by switching on the SCR for longer or shorter periods of time (see section 19).

Fig.17.7 shows an SCR triggered by a train of pulses. The SCR switches on at the positive edge of the pulse and stays on until the input falls to zero, producing an output which is part of the positive half cycle of the input as shown.

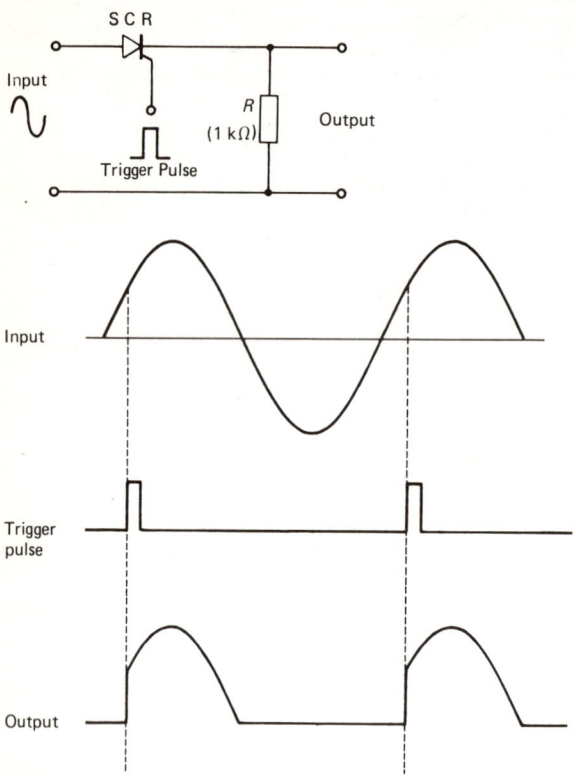

Fig.17.7 Pulse triggered SCR

Fig.17.8 shows a simple circuit using a variable resistor VR_1 to control the SCR. The SCR is triggered by the
input waveform itself. When VR_1 is set to its minimum
value, triggering occurs early as shown in 17.8a. As VR_1
is increased, the triggering is delayed since the amplitude of the waveform going into the gate gets smaller. At
the maximum value of VR_1 the SCR is triggered just before
the positive peak as shown in 17.8b. Note that the SCR
in the circuit under consideration may only be triggered
during the first quarter of the cycle, i.e. before the
positive peak arrives at the gate. If it fails to do
that, it will not trigger at all and the output would be
zero.

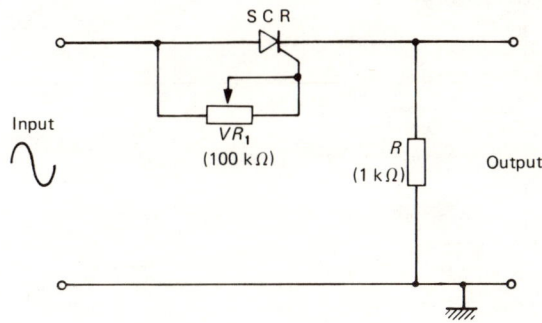

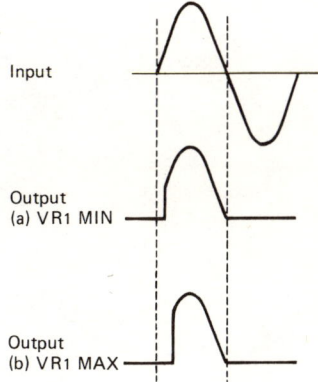

Fig.17.8 The SCR rectifier

To achieve triggering during the second quarter of the cycle, i.e. after the positive peak has passed, a phase shift network is used as shown in Fig.17.9. Capacitor C and resistor VR_1 act as the phase shift network. The waveform fed into the gate suffers a delay (or a phase shift) as shown in 17.9b. As stated above, the SCR may only be triggered before the positive peak arrives at the gate. But due to the phase shift, by the time the positive peak arrives at the gate, the positive peak of the input has already passed. Hence the SCR may now be triggered during the second quarter of the input cycle, as shown in 17.9c.

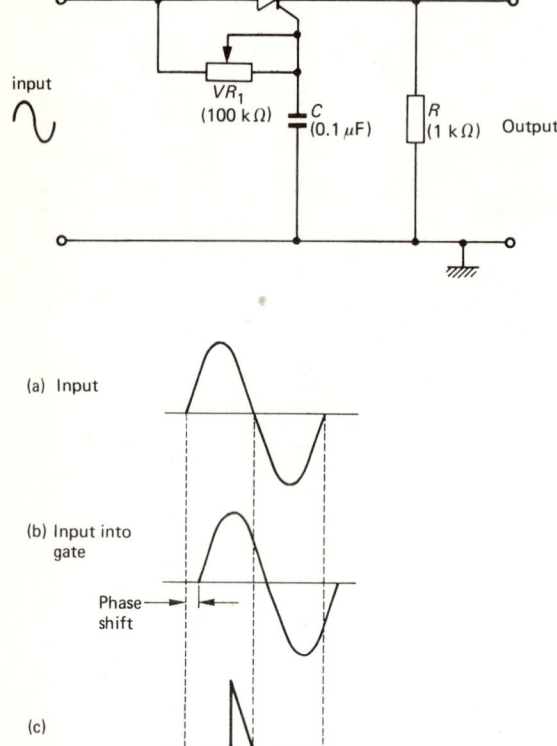

Fig.17.9 *SCR rectifier employing a phase shift network VR_1-C*

The Varactor

It is found that a reverse biased pn junction diode has a small capacitance across it which varies with its reverse voltage. This fact is used in the fabrication of integrated circuits in order to include capacitors within the silicon slice.

Reverse biased diodes are used as variable-capacitor diodes known as vari-cap diodes or varactors (Fig.17.10). Varactors are employed in, among other things, automatic frequency control systems (see p.156). What is commonly known as electronic tuning also employs varactors as the tuning capacitors. Preset channel selection is achieved by using a potentiometer which fixes the reverse d.c. bias across the varactor, determining the tuning capacitance. A swing of approximately 30 V is sufficient to tune the circuit through all the u.h.f. channels.

Compared with the common variable capacitor, the varactor is small in size, more sensitive, and has very high stability and reliability.

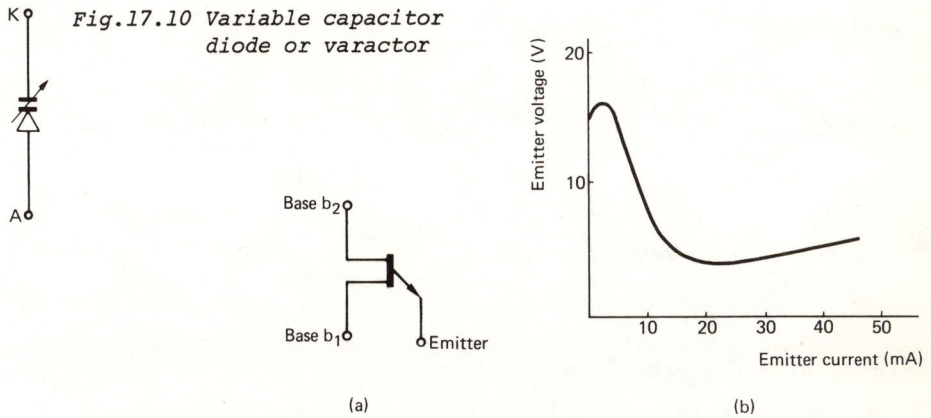

Fig.17.10 Variable capacitor
diode or varactor

(a)

(b)

Fig.17.11 Unijunction
(a) p-type symbol (b) characteristics

The Unijunction

Another device which has a negative resistance is the unijunction. Fig.17.11 shows the symbol and characteristics of a p-type unijunction. As soon as the emitter voltage is high enough to forward bias the pn junction between the emitter and base b_1, current begins to flow from the emitter with the voltage across the junction falling to a low value (about 0.6 V). Unijunctions are often used as oscillators (see p.144) and for triggering silicon controlled rectifiers.

18 R-C Filters, Clippers, and Clampers

A filter allows one frequency band to pass through without attenuation (weakening) and cuts off all other frequencies. The frequency at which cut-off occurs is known as the cut-off frequency f_c (Fig.18.1).

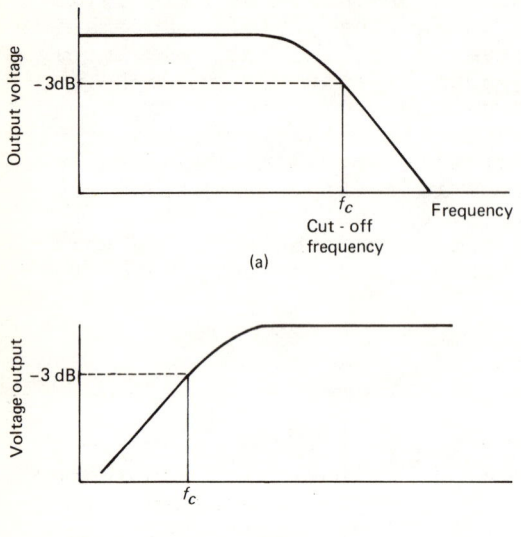

(a)

(b)

Fig.18.1 Frequency response
(a) Low pass filter   (b) Hi-pass filter

Effect on a Square Wave

A square wave is a complex waveform consisting of a fundamental and an infinite number of odd harmonics.* A square wave can thus be thought of as consisting of a low frequency component represented by the flat top and bottom, and a high frequency component represented by the fast rising and falling edges.

When fed through a filter, the square wave suffers a distortion in its shape. In general, a low pass filter will distort the high frequency edges, rendering them less sharp and producing rounded corners as shown in Fig.18.7b. (A low pass filter has the same effect on a square wave as an amplifier with insufficient bandwidth.) Conversely, a high pass filter distorts the flat top and bottom of a square wave as shown in Fig.18.5b.

*See the companion book *Principles and Systems for Radio and TV Mechanics.*

R-C Filters

The simplest type of a filter is an R-C filter. It uses the fact that the reactance of a capacitor varies inversely with frequency, while a resistance remains constant as frequency changes. Fig.18.2 shows a capacitor in series with a resistance. At low input frequencies, the reactance of the capacitor is very large compared with resistance R. Hence the voltage v_c across the capacitor is large while that across the resistor v_r is low. At high input frequencies the opposite is true, namely v_c is low and v_r

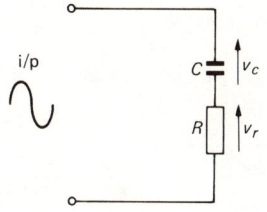

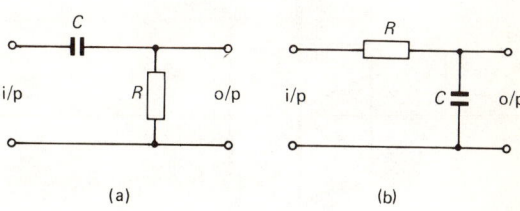

Fig.18.2

Fig.18.3 (a) R-C hi-pass filter
(b) R-C low pass filter

is high. Therefore if an output is taken across the capacitor (Fig.18.3a), low frequencies will predominate while high frequencies will be greatly attenuated. In other words we have a low pass filter. Conversely, an output across the resistor produces a hi-pass filter (Fig. 18.3b). The values of R and C determine the cut-off frequency.

The Differentiator

The differentiator is a hi-pass filter which when fed with a square wave produces high frequency spikes. Fig.18.4

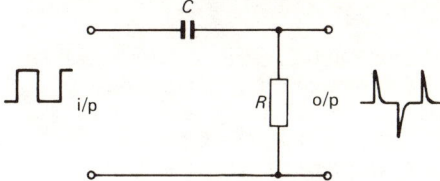

Fig.18.4 R-C differentiator

shows an R-C differentiator. Capacitor C allows the high frequency component AB through. The capacitor then begins to charge up to 10 V. Provided the time constant (the product of C and R) is short compared with the periodic time of the input waveform, the capacitor will become

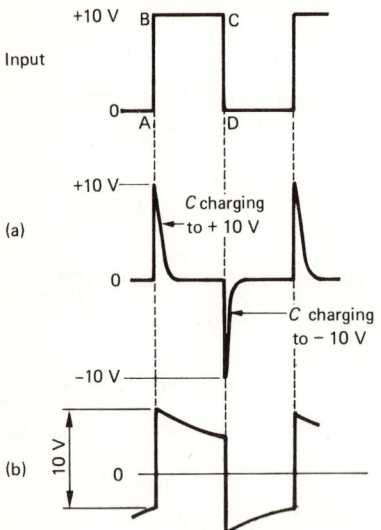

Input

+10 V

B   C

0   A   D

(a)

+10 V

C charging
to + 10 V

0

C charging
to − 10 V

−10 V

(b)

10 V

0

Fig.18.5 Output from the differentiator in Fig.18.4
        (a) RC time constant very short compared with period of input
        (b) RC time constant comparable with period of input

fully charged to 10 V well before the next high frequency
component CD arrives (Fig.18.5a). When the capacitor is
fully charged, the voltage across the resistor, i.e. the
output, is zero. The falling edge CD represents a 10 V
drop and, being a high frequency component, will pass
through the capacitor bringing the output voltage down to
−10 V. The capacitor then begins to charge up to −10 V,
and given the short time constant the capacitor will be
fully charged, with the output falling to zero well before
the next high frequency edge arrives and so on. The
effect of a time constant that is long compared with the
period of the input is shown in Fig.18.5b.

The Integrator

The integrator is a low pass filter which when fed with a
square wave produces a triangular waveform. Fig.18.6
shows an *R-C* integrator. At the first rising edge of the
input (Fig.18.7), the capacitor begins to charge up
towards +10 V. Provided the time constant *CR* is long
compared with the period of input, the falling edge CD
arrives well before the capacitor is fully charged (Fig.
18.7a). The falling edge CD will attempt to charge the
capacitor in the negative direction. Again, due to the
long time constant, before the capacitor can charge up
negatively, the next edge EF arrives, and so on. The
output waveform will, therefore, be triangular in shape
having a low amplitude compared with the input.

   The effect of a short time constant is shown in Fig.
18.7b. Note that for both a differentiator and an integrator, the time constant is always compared with the

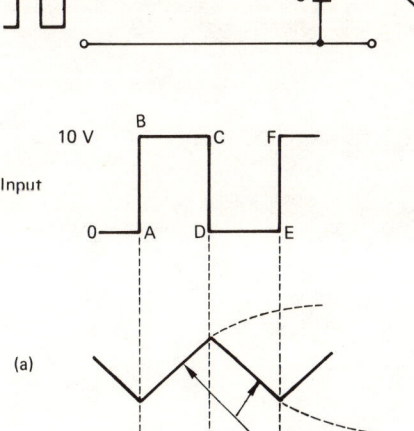

Fig.18.6 R-C integrator

Fig.18.7 Output from integrator in Fig.18.6
  (a) RC time constant very long compared with period of input
  (b) RC time constant comparable with period of input

period of the input. A time constant of 100 μs for example is long compared with a periodic time of say 5 μs (input frequency of 200 kHz) but short compared with a periodic time of say 5 ms (an input frequency of 200 Hz).

Diode Clipping

A CLIPPER, also known as a limiter, flattens or slices the top or bottom or both of a waveform. The circuit in Fig. 18.8 clips the whole of the negative half cycle. The forward voltage drop of the diode would produce a 0.6 V negative excursion shown in dotted line. Usually the forward voltage drop is neglected and diodes are assumed ideal.

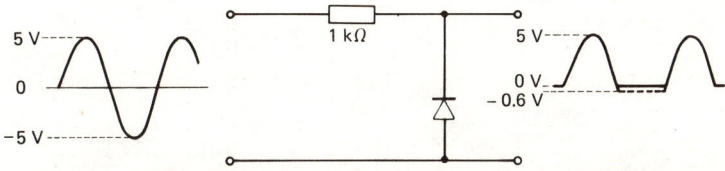

Fig.18.8

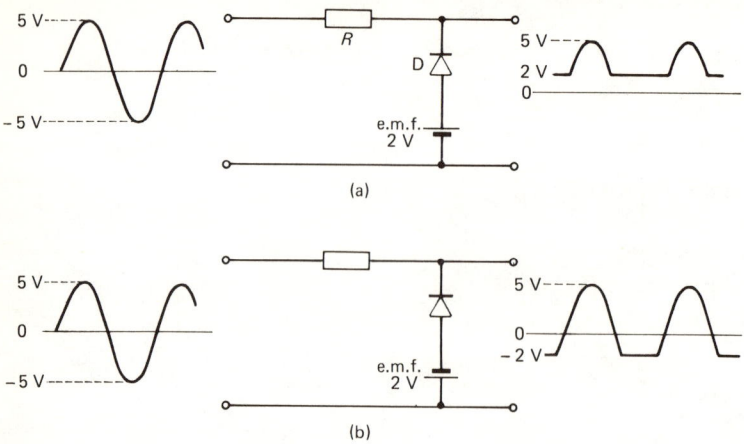

Fig.18.9 (a) Clipping at + 2 V
         (b) Clipping at - 2 V

To produce clipping at other levels, an e.m.f. may be added in series with the diode to forward bias the diode (Fig.18.9a) or to reverse bias the diode (18.9b). In the circuit in 18.9b the diode conducts only when its cathode is at or below -2 V, assuming ideal diode. Cathode voltages above -2 V will keep the diode reverse biased, i.e. open circuit, and hence that part of the waveform appears at the output.

A simple method of working out the output of a diode clipper with an external e.m.f. is as follows:
(1) Draw a line representing the e.m.f. level along the input waveform.
(2) If the e.m.f. is connected in such a way as to forward bias the diode, then the larger part of the waveform will be clipped off and vice versa.

The circuits in Fig.18.10 produce clipping of both positive and negative half cycles. In 18.10a, D_1 clips the positive half while D_2 clips the negative half cycle. For ideal diodes the output would be zero. However, due to the forward voltage drop across the diodes (0.6 V for silicon diodes) clipping occurs at +0.6 V and -0.6 V.

Fig.18.10b shows a circuit which is in effect a doubling of that shown in 18.9b. D_1 clips the positive half at +2 V while D_2 clips the negative half at -4 V.

Zener Clipping

Zener diodes can also be used for clipping a waveform as shown in Fig.18.11. Z_1 in 18.11b conducts for the whole positive half cycle. However Z_2 stays off until the input voltage exceeds its breakdown voltage (6 V in this case) clipping the input as shown. During the negative half cycle, Z_1 conducts and Z_2 clips the waveform at -9 V.

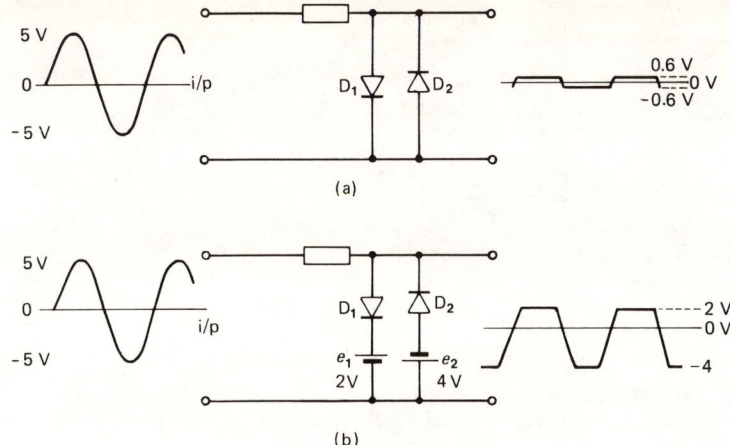

Fig.18.10 *Positive and negative clipping*

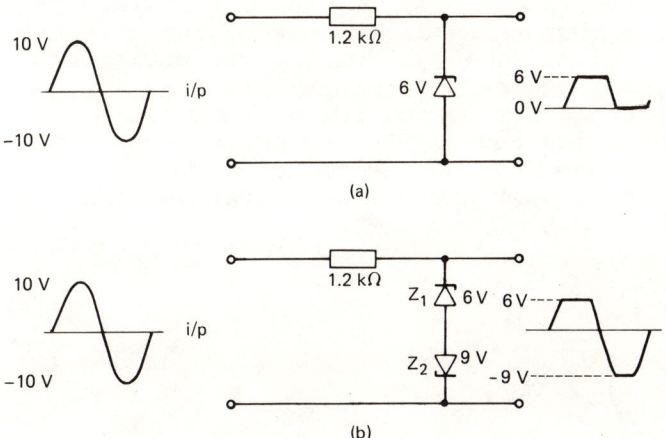

Fig.18.11 *Zener clipping*

Transistor Clipping

As explained in section 10, clipping can be achieved by overloading an amplifier. The circuit in Fig.18.12 produces an output which is practically a square wave. The transistor has no bias voltage, hence it is at cut-off without an input. During the positive half cycle of the input, the transistor remains off and the output is at $-V_{cc}$. During the negative half cycle the transistor switches on (its base going away from the emitter) and

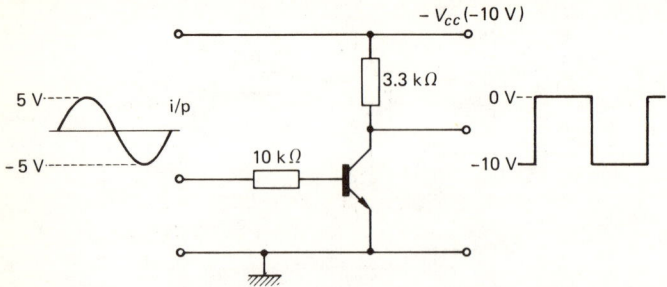

Fig.18.12 Transistor clipping

provided the input is large enough it saturates, giving
an input of zero. The output is a square wave and hence
the circuit is known as a squarer.

Another squarer circuit is the emitter coupled pair
shown in Fig.18.13. TR_1 and TR_2 are coupled together
with a common emitter resistor R_5. Chains R_1–R_2 and
R_6–R_7 provide the forward bias for TR_1 and TR_2 respectively. The current through R_5 is the sum of the currents
through TR_1 and TR_2, i.e. $I_1 + I_2$. During the positive
half of the input, TR_1 begins to conduct more than TR_2,
V_e goes up and TR_2 begins to go towards cut-off (its
emitter going towards its base). As the current through
TR_2 decreases so V_e goes down making TR_1 conduct even
more, and so on. The result of this regenerative action

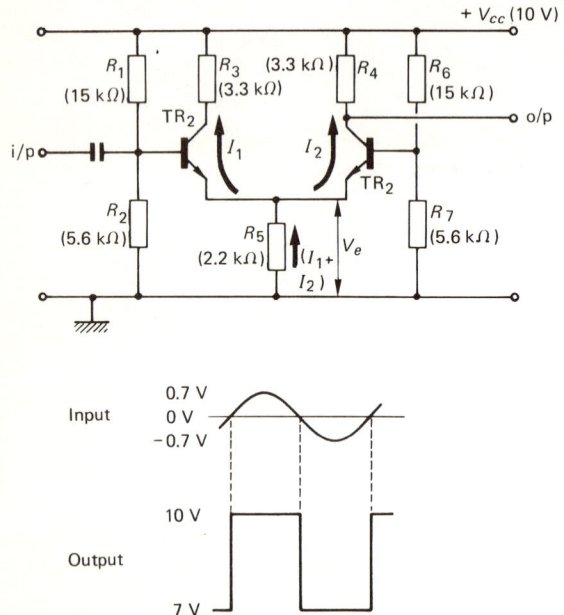

Fig.18.13 Emitter coupled clipping circuit

is that TR_1 saturates and TR_2 turns off very quickly and remains in that state till the negative half cycle arrives. Thus during the positive half cycle the output taken at the collector of TR_2 is at V_{cc}.

During the negative half cycle, TR_1 is turned off and TR_2 saturates, bringing the output down to the emitter voltage V_e. The circuit in Fig.18.13 is more sensitive than that using a single transistor. A square wave will be produced at the output provided the peak input is greater than the transistor forward bias.

Schmitt Trigger

A very useful circuit for squaring, clipping and pulse shaping is the Schmitt trigger shown in Fig.18.14. With no input, TR_1 base is at zero potential and the transistor is cut-off. TR_2 however is saturated since its base has a positive voltage determined by potential divider R_2-R_4-R_5. The output (TR_2 collector) is practically at zero potential. TR_2 current flowing through the common emitter resistor R_3 gives the emitters a positive voltage which reverse biases TR_1.

When the input is increased in the positive direction, TR_1 begins to conduct when its base voltage reaches that of TR_2 base. In the same way as in the common emitter circuit discussed previously, a regenerative action takes place, resulting in TR_1 saturating and TR_2 turning off very quickly, with the output going to V_{cc}.

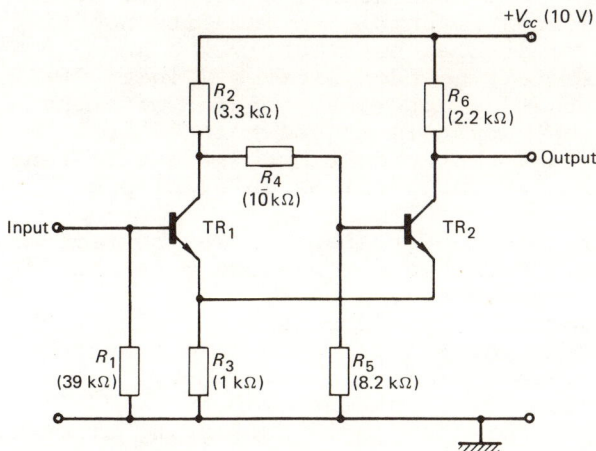

Fig.18.14 Schmitt trigger circuit

The Clamper

Also known as d.c. restorer, the clamper provides a.c.
waveforms with a d.c. level. A clamper does not change
the shape of the input signal, it just provides it with a
d.c. component.

There are many occasions where it is important to
retain the d.c. level of a signal. For example, a video
signal contains a d.c. level which corresponds to the
average brightness of the scene. This d.c. level is
sometimes lost due to a.c. coupling and has to be restored.
Clamping is also used to provide Class C biasing for transistors when used in oscillators for example (see section
20). Fig.18.15 shows a clamper circuit.

Falling edge AB (Fig.18.16) of the input represents a
drop of 10 V which, being of high frequency, will pass
through the capacitor making point X (the output) go
negative. The diode is forward biased and assuming it to
be ideal, it will act as a short circuit across the output
clamping point X to zero. The capacitor quickly charges
up through the small diode forward resistance in the negative direction as shown in Fig.18.15b. This charge is
retained till the next rising edge arrives.

Rising edge CD represents a 10 V positive step which,
being of high frequency, passes through the capacitor
shifting point X from zero volts to + 10 V (Fig.18.15b).
The diode is now reverse biased (i.e. open circuit) with
10 V across resistor R. Current begins to flow and
attempts to charge the capacitor in the opposite direction;
the voltage at the output begins to fall. However, if the
time constant CR is long compared with the periodic time
of the input, the capacitor is unable to lose its negative
charge and the output remains at practically the same
potential till the next falling edge arrives (Fig.18.16a).

For a time constant that is not long compared with the
period of the input, the output is as shown in 18.16b.
With a short time constant the circuit becomes a differentiator, having its negative going spikes fully clipped
(Fig.18.16c).

To ensure a long time constant resistor R may be removed,
making the time constant equal to $C \times r$ where r is the
reverse resistance of the diode.

The diode may be reversed to produce zero clamping with
negative-going waveform (Fig.18.17). Note here that the
input itself has a d.c. level which is removed by the
capacitor and hence does not affect the output.

To draw the output waveform for a given clamping circuit,
first redraw the input waveform and if the diode is connected in such a way as to clip positive excursions, the
waveform will be negative-going only and vice versa,
having exactly the same amplitude as the input.

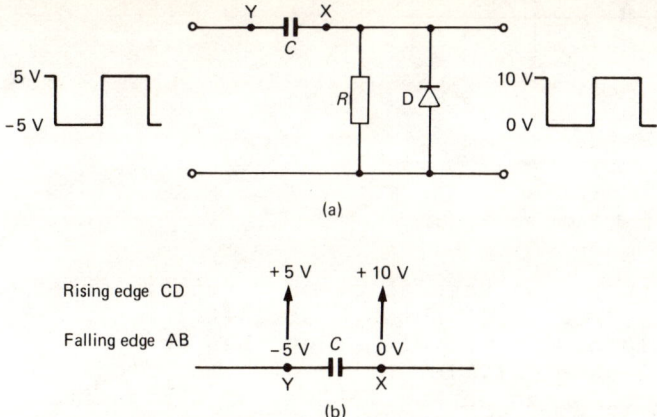

Fig.18.15 Clamping
        (a) Circuit producing zero clamping
        (b) Output (point X) changing from zero to 10 V
        as the input changes from - 5 V to + 5 V

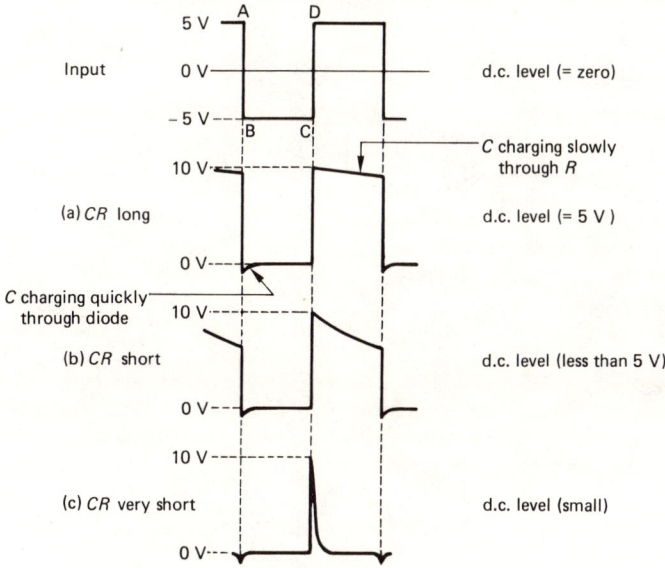

Fig.18.16 Output of clamper in Fig.18.14: (a) with time constant very
        long, (b) with time constant short and (c) with time
        constant very short

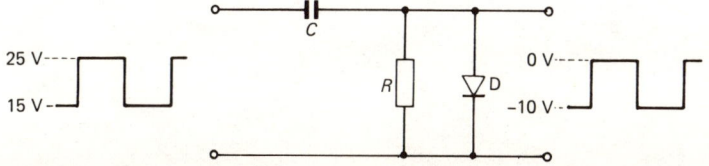

Fig.18.17 Zero clamping with negative going output

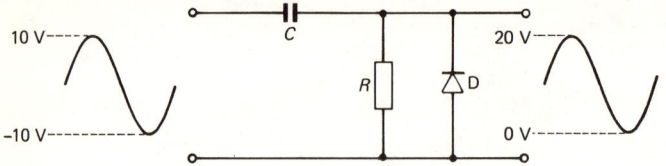

Fig.18.18 Clamping of a sine wave

Clamping circuits can be used to clamp waveforms other than square waves. Fig.18.18 shows a circuit for sine wave clamping. Clamping at levels other than zero are possible by adding an e.m.f. in series with the diode to give it a forward bias, as in Fig.18.19a, or a negative bias as in 18.19b. In 18.19b the diode is given a reverse bias of 5 V thus preventing the output from going above +5 V. Since a clamping retains the amplitude of the input, the output drops to -15 V resulting in a total amplitude of 20 V.

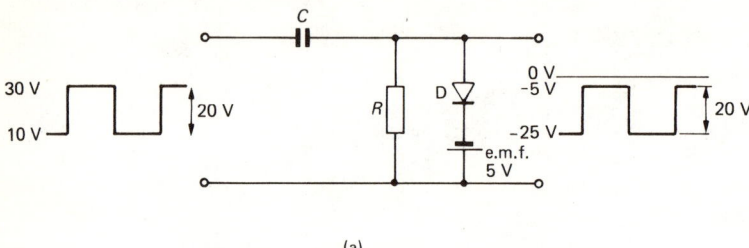

(a)

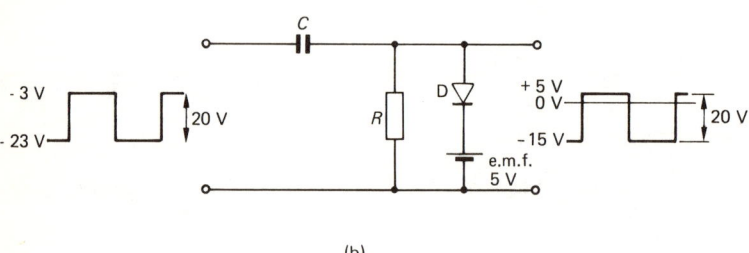

(b)

Fig.18.19 (a) Clamping at - 5 V
(b) Clamping at + 5 V

19 D.C. Power Supplies

Rectification

There are two types of rectifier.

(1) HALF-WAVE RECTIFIER, shown in Fig.19.1. Diode D_1 in 19.1a allows only the positive half cycles through, producing a positive-going output. When the diode is reversed as in 19.1b negative-going half cycles are produced. The output waveform contains a d.c. level which is approximately one-third of the peak voltage (Fig.19.2).

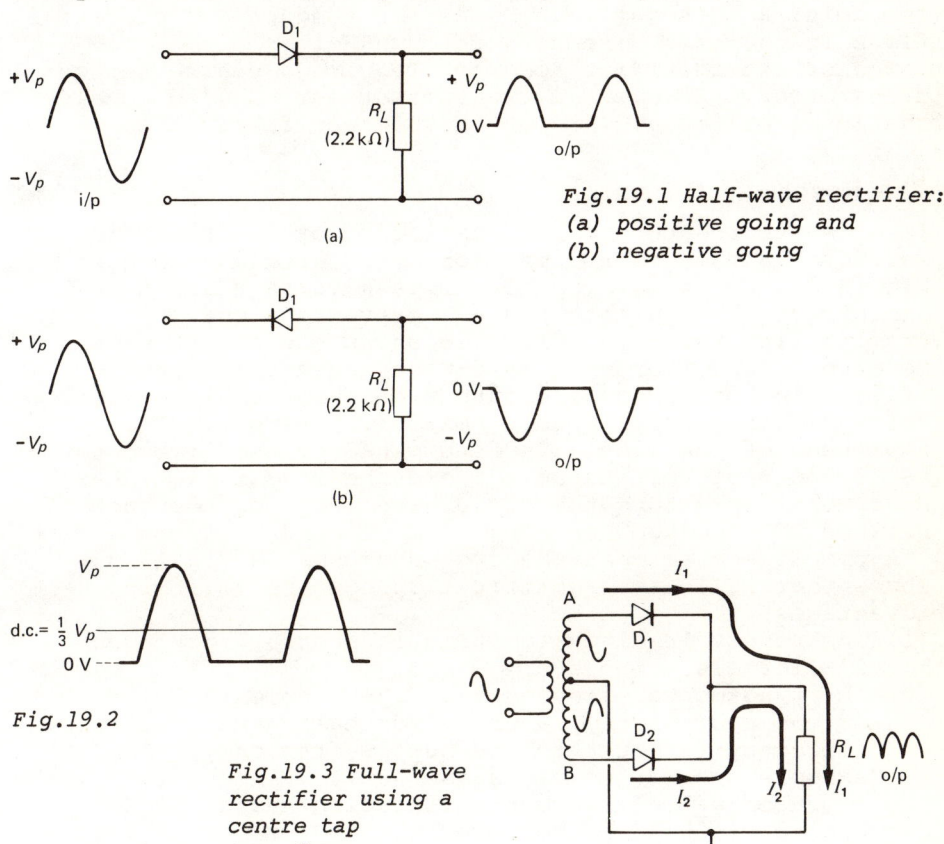

Fig.19.1 Half-wave rectifier:
(a) positive going and
(b) negative going

Fig.19.2

Fig.19.3 Full-wave
rectifier using a
centre tap
transformer

(2) FULL-WAVE RECTIFIER, shown in Fig.19.3. A transformer with a centre tapped secondary is used. Two equal and opposite waveforms are developed across each half of the secondary winding. For one half of the input, point A is positive and point B is negative with respect to the centre tap, and vice versa for the other half cycle. When A is positive, D_1 conducts and current I_1 flows as shown. When B is positive, D_2 conducts and current I_2 flows through R_L in the same direction as I_1. The output

waveform is therefore as shown in Fig.19.4 with a d.c.
level double that of the half-wave rectifier at approximately two-thirds of the peak voltage.

The Bridge Rectifier

Another circuit that produces full-wave rectification is
that shown in Fig.19.5 and known as the bridge rectifier.
During the positive half cycle of the input (Fig.19.6a),
point A is positive and point B is negative. D_1 and D_2
conduct and current I_1 flows downwards through load R_L.
During the negative half cycle (19.6b) point B is positive
and point A is negative. D_2 and D_4 conduct and current I_2
flows in the same direction through R_L.
 The bridge rectifier does not require a centre tapped
transformer. However, a transformer may be used to alter
the level of the a.c. input to the rectifier.

The Reservoir Capacitor

In order to reduce the a.c. variation of the rectified
waveform, a reservoir capacitor C_1 is used as shown in
Fig.19.7. C_1 charges up to the peak voltage and then
discharges through the load R_L, preventing the voltage
from falling rapidly. Fig.19.8 shows the effect of the
reservoir capacitor on a half-wave and a full-wave rectifier. In both cases the output contains a higher d.c.
voltage with a superimposed small a.c. ripple. The
amplitude of the ripple is determined by the time constant
CR of the reservoir capacitor and the load resistance.
Reservoir capacitors therefore have large capacitance of
between 100 and 5000 µF.
 A comparison between the two waveforms in Fig.19.8
shows that full-wave rectification has the following
advantages:
(1) The reservoir discharge time is shorter, giving a
    smaller a.c. ripple.
(2) The fundamental frequency of the ripple is twice that
    of the a.c. supply, whereas for half-wave, the ripple
    frequency is equal to the supply frequency. For
    example, using the mains as the a.c. input, the ripple
    produced by full-wave rectification has a frequency
    $2 \times 50 = 100$ Hz as compared with 50 Hz for half-wave
    rectification. Full-wave ripple is easier to filter
    out due to its higher frequency as will be seen later.

No-load Voltage

The no-load voltage is the terminal voltage of the power
unit when load current is zero, i.e. when the load is
removed. Fig.19.9 shows a simple power supply with no-
load resistance. Reservoir capacitor C charges up to the

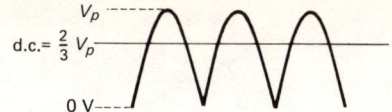

Fig.19.4 D.C. level of a full-wave rectified sine wave is $2/3$ of the peak

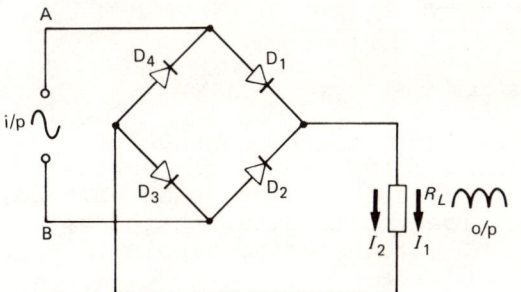

Fig.19.5 Bridge rectifier

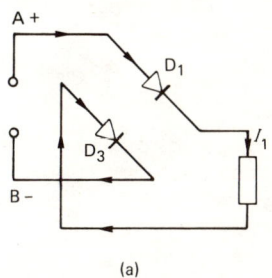

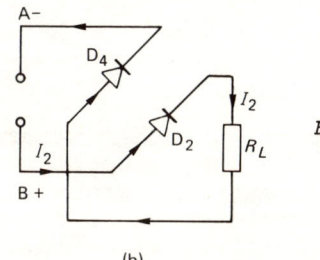

Fig.19.6

(a)   (b)

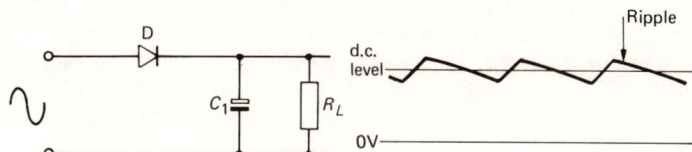

Fig.19.7 D.C. power supply with reservoir capacitor C

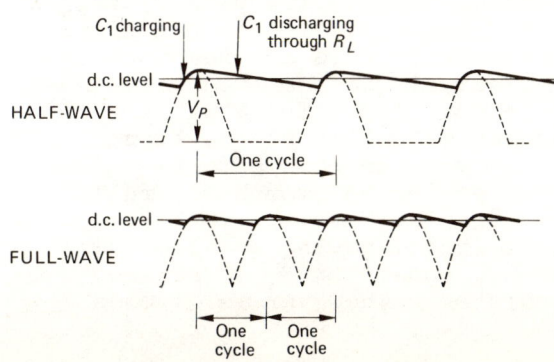

Fig.19.8 The effect of the reservoir capacitor on a rectified sine wave

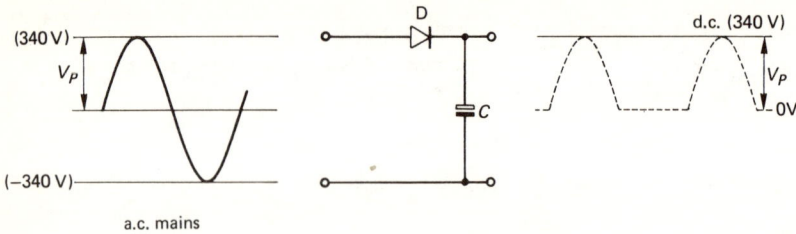

Fig.19.9 No load voltage = peak input voltage

peak voltage as normal. However with the absence of a
load resistor, the capacitor retains the charge across it,
thus producing a steady output (equal to the peak voltage)
without any ripple. The no-load voltage is, therefore,
the highest terminal voltage. If the mains supply is
used, the no-load terminal voltage

$$V_{n\ell} = \text{peak voltage} = 1.4 \times \text{r.m.s. voltage}$$

$$= 1.4 \times 240 = 340 \text{ V}$$

Peak Inverse Voltage

An important factor that must be considered in choosing
the diodes for d.c. power supplies is the maximum voltage
the device has to sustain across it during the non-
conducting half cycle known as the peak inverse voltage
PIV. Consider the circuit in Fig.19.9. The highest
potential at the cathode of diode D is the no-load voltage
340 V. The anode potential varies between the positive
peak of +340 V and a negative peak of -340 V. The maximum peak voltage that the diode has to sustain is when
the anode is at the negative peak of -340 V, giving a PIV
of 340 + 340 = 680 V. It can therefore be seen that the
PIV is 2 × no-load voltage, i.e. 2 × the peak voltage of
the a.c. input.

R-C Smoothing

The a.c. ripple may be reduced using a low pass filter or
smoothing circuit. R_1 and C_2 in Fig.19.10 form a simple
R-C filter. To provide adequate ripple attenuation, the
time constant R_1C_2 must be very large compared with the
periodic time of the ripple waveform. For a given time
constant, the shorter the periodic time of the ripple
(i.e. the higher its frequency), the more effective is
the smoothing. Hence it is easier to smooth out full-
wave ripple than half-wave ripple.

Smoothing capacitor C_2 is large, having a value comparable to that of the reservoir capacitor C_1 of between 100
to 5000 µF. Smoothing resistor R_1 however is given a

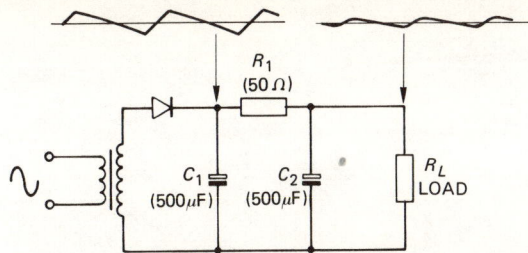

Fig.19.10 D.C. power supply with R-C filter R_1-C_2

small value, otherwise a large d.c. drop will develop
across it reducing the terminal voltage. Values of
smoothing resistors are in the region of 1 to 100 Ω
depending on the load current.

L-C Smoothing

A more effective smoothing circuit is shown in Fig.19.11.
L_1 and C_2 form a low pass filter. Smoothing choke L_1 has
a large inductance (100 mH - 10 H) which tends to smooth

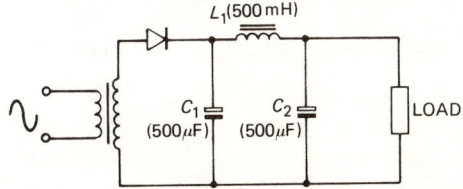

Fig.19.11 D.C. power supply with L-C filter L_1-C_2

out variations in current through it, reducing further
the a.c. ripple at the output. Its low resistance is a
further advantage while its size is a disadvantage. Note
that while the reservoir capacitor C_1 increases the d.c.
level of the power unit, the smoothing circuit leaves the
d.c. level virtually unchanged. It only attenuates the
ripple content.

Regulation

An increase in the current supplied by a power unit causes
the terminal voltage to drop. This is due to its internal
resistance which is the sum total of the resistance of the
transformer winding, diodes, smoothing resistor or choke.
Fig.19.12 shows the change in the terminal voltage of a
d.c. supply as load current is increased. This curve is
known as the regulation curve. The terminal voltage is

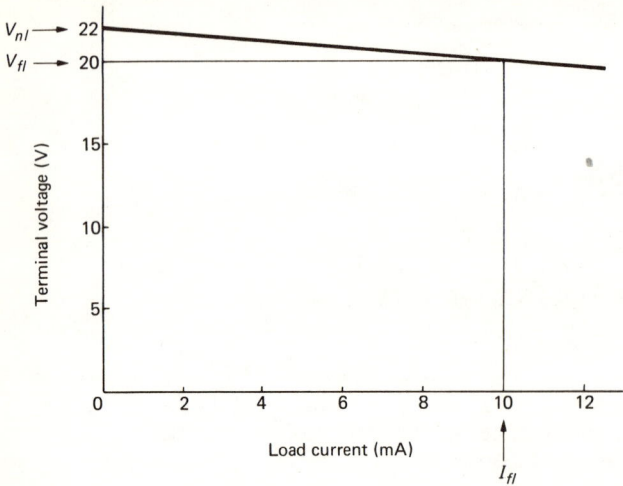

Fig.19.12 Regulation curve for an unstabilised power supply

at a maximum when the load current is zero, i.e. at no
load. When it is supplying the specified full load
current, the terminal voltage is referred to as the full
load voltage.
   Regulation is given as a percentage and it is defined
as

$$\frac{\text{No load voltage} - \text{Full load voltage}}{\text{Full load voltage}} \times 100\%$$

$$= \frac{V_{n\ell} - V_{f\ell}}{V_{f\ell}} \times 100\%$$

For example if a d.c. supply provides 22 V when on no
load and 20 V when supplying a full load current of 10 mA,
its regulation is

$$\frac{V_{n\ell} - V_{f\ell}}{V_{f\ell}} \times 100\% = \frac{22 - 20}{20} \times 100\% = \frac{2}{20} \times 100\% = 10\%$$

Stabilised Power Supplies

The regulation of a power supply can be improved by the
use of voltage stabilisation or regulation. The terminal
voltage may in this way be maintained at a constant level
as the load current is varied. There are two types of
stabilisation: the shunt and the series.

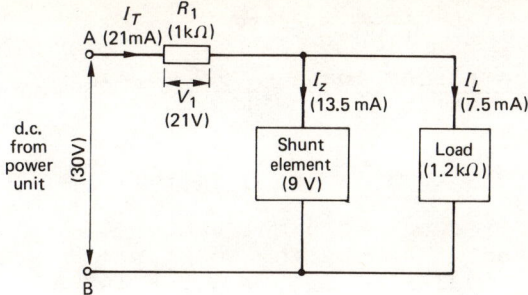

Fig.19.13 Block diagram for a shunt stabiliser

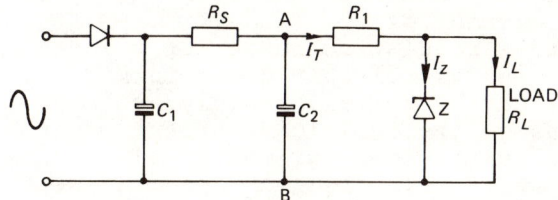

Fig.19.14 A shunt stabilised power supply

Shunt Stabilisers

Fig.19.13 shows a block diagram for a shunt stabiliser. Fig.19.14 shows a circuit for a shunt stabilised power supply in which the stabiliser element is the zener diode. The zener operates at its breakdown voltage, keeping the voltage across it and hence the output voltage constant for a very large variation of current through it. A gas-filled diode may also be used instead of the zener with the same results.

Shunt stabilisation is based on a current-sharing principle whereby the sum of the load current I_L and the zener current I_z is kept constant. If the load current is increased by say 2 mA, then the shunt stabiliser current is decreased by the same amount of 2 mA and vice versa.

The series dropper resistor R_1 takes the total current, thus developing a voltage V_1 across it equal to the difference between the unregulated d.c. voltage V_{AB} and the zener breakdown voltage V_z.

$$V_1 = V_{AB} - V_z$$

Using the values given in Fig.19.13, $V_0 = V_z = 9$ V.

Voltage across R_1, $V_1 = V_{AB} - V_z = 30 - 9 = 21$ V.

Therefore, total current $I_T = \dfrac{21 \text{ V}}{1 \text{ k}\Omega} = 21$ mA

Load current $I_L = \dfrac{\text{Voltage across load}}{\text{Load resistance}} = \dfrac{9 \text{ V}}{1.2 \text{ k}\Omega} = 7.5$ mA

Therefore

Zener current $I_z = I_T - I_L = 21 - 7.5 = 13.5$ mA

If the load current is now reduced to 5 mA, a drop of 7.5 - 5 = 2.5 mA, the current taken by the zener will increase by 2.5 mA to a new value of 13.5 + 2.5 = 16 mA.

Under no-load condition when $I_L = 0$, the whole of the total current I_T will now be taken by the zener: $I_T = I_z$. This is one disadvantage of the shunt stabiliser, namely that regardless of the load, the power supply has to provide a constant maximum current of I_T.

Fig.19.15 shows a typical regulation curve for a shunt stabilised power supply, for the circuit given in Fig. 19.14. The load voltage begins to fall rapidly as the load current increases above the full load current of just under 21 mA. At this value the load current is almost equal to the total current I_T. The fall in the voltage occurs when the zener current is too small for breakdown to occur. For good regulation, the unregulated d.c. voltage is normally chosen to be three times the breakdown voltage of the zener.

Series Stabilisation

A better and more efficient stabiliser is the series stabiliser, which employs a transistor or a thyristor as the series element. Fig.19.16 shows a simple block diagram for a series stabiliser. It consists of a series control element with a bleed resistor R to provide the minimum load current.

Series Transistor Stabiliser

A basic circuit for a series stabiliser employing a transistor is shown in Fig.19.17. The output is taken at the emitter of the TR_1 and, as can be seen from Fig.19.18 where the same circuit is redrawn, TR_1 is used as an emitter follower. The zener maintains the base of the transistor at a constant potential. The emitter follows the base potential so that it is 0.6 V (for silicon) below the base voltage for a normal forward bias, thus keeping the output voltage constant.

The emitter follower acts as a current amplifier which helps to supply large load currents for a small input current. The zener acting as a regulator as well as a voltage reference diode takes a smaller current compared with the zener used in the shunt stabiliser. For good regulation the zener current should be approximately 5 times the base current.

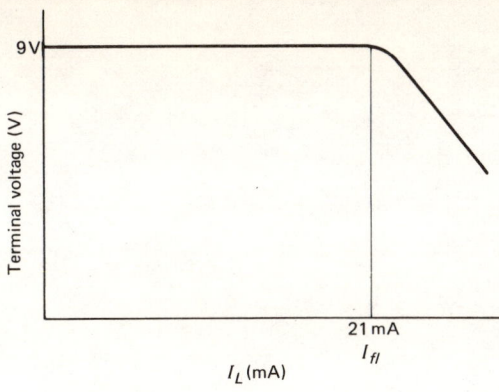

Fig.19.15 Regulation curve for a
stabilised power supply

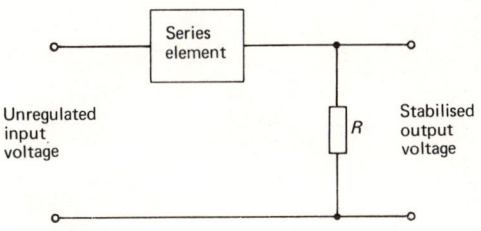

Fig.19.16 Block diagram for a
series stabilised power
supply

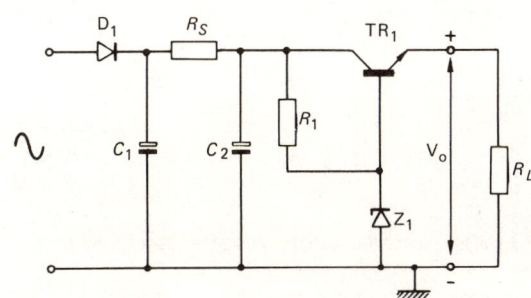

Fig.19.17 A series stabilised
power supply

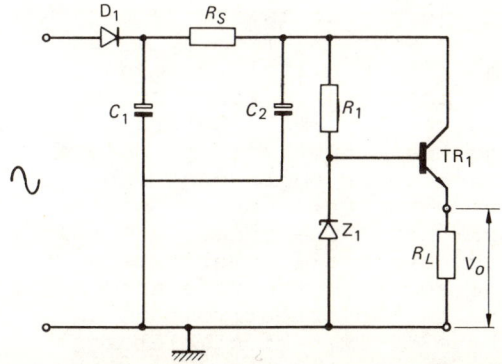

Fig.19.18 Circuit in Fig.19.17
redrawn to show the
common emitter config-
uration of TR₁

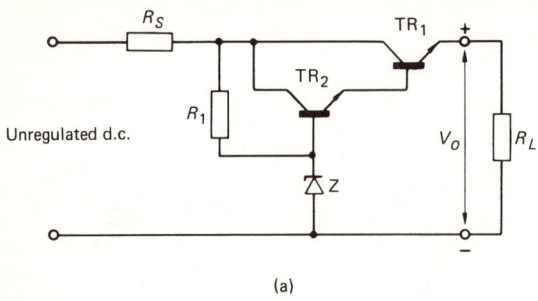

(a)

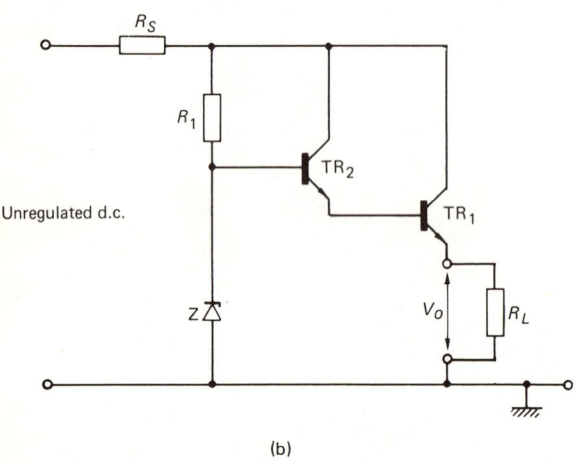

(b)

*Fig.19.19 Series stabilised power supply using two emitter followers.
(The circuit is redrawn in (b))*

The simple series stabiliser considered above suffers from two main disadvantages:
(1) For large load currents, power zeners and high gain transistors are required.
(2) Limited sensitivity.
The first may be overcome by increasing the current gain by means of an additional transistor TR_2 forming a second emitter follower stage as shown in Fig.19.19. Load current may be very large (in amps) while zener current remains very small. Sensitivity, i.e. voltage stability, may be improved by amplifying the output voltage before comparing it with the reference zener voltage as shown in Fig.19.20. TR_1 is the normal series transistor, while TR_2 is the error voltage amplifier. The zener is a reference diode only and hence may have small power rating.
TR_2 compares the output voltage with the reference voltage of the zener. Change in the output voltage is amplified and fed into the base of TR_1 which maintains the output voltage constant. For example if due to some external disturbance the output voltage V_O increases, TR_2

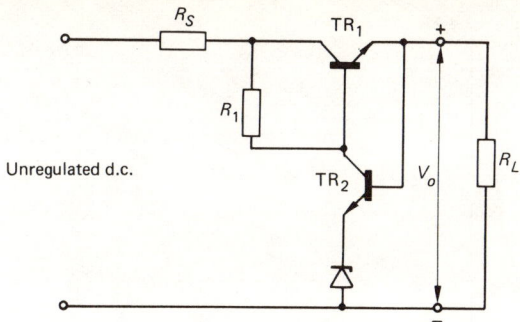

Unregulated d.c.

Fig.19.20 Series stabiliser using error amplifier TR₂ to improve its sensitivity

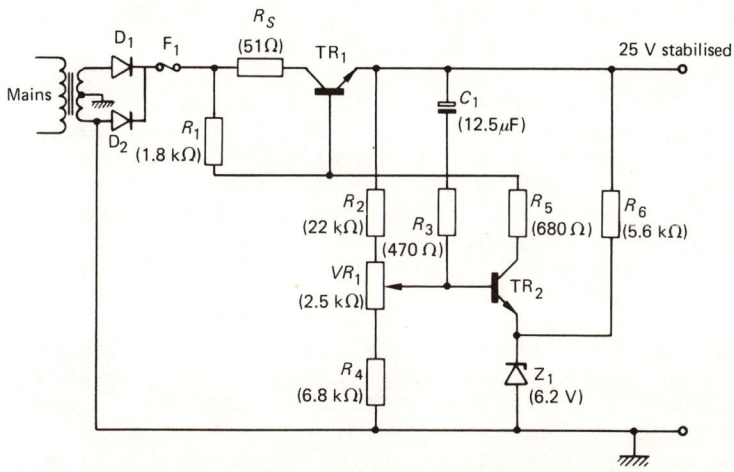

Fig.19.21 25 V stabilised power supply used in TCE 8000 colour TV chassis. (By permission of Thorn Consumer Electronics)

base goes up away from the emitter which is at the fixed voltage of the zener, current through TR_2 increases and its collector voltage decreases. TR_1 base potential thus suffers a drop towards the emitter which reduces the current through TR_1, making load voltage V_O go down and thereby compensating for the original increase.

Many modifications may be made to the series stabiliser in order to improve its stability and regulation. Fig. 19.21 shows a practical circuit used to produce a stabilised 25 V for a colour TV receiver. TR_1 is the normal series stabiliser. TR_2 is a voltage amplifier with its base voltage adjustable by VR_1 to enable the 25 V output to be set up. C_1 and R_3 provide what is known as electronic smoothing. The 100 Hz ripple on the output is fed through TR_2 back to the base of TR_1 in anti-phase to the ripple present there, thus cancelling it out.

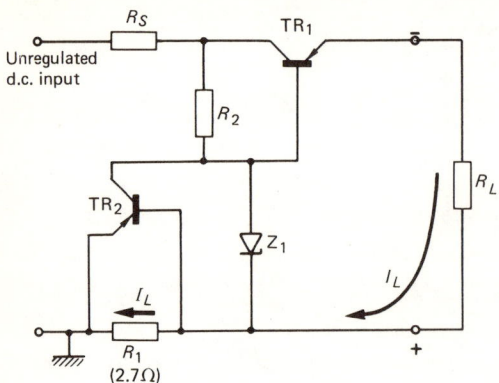

Fig.19.22 Series stabilised power supply employing overload protection transistor TR$_2$

OVERLOAD PROTECTION CIRCUIT One of the problems encountered by series stabiliser is the protection of the series transistor. An excessive current through the transistor due to an overload or a short circuit may result in permanent damage. One method of protection is shown in Fig. 19.22. TR$_2$ is the overload protection transistor. R_1 falling in the return path of load current I_L develops a p.d. across it which provides the transistor with a b-e voltage. Provided the load current is normal, the voltage across R_1 is small, keeping TR$_2$ off. If the load current is increased above its normal value, the potential across R_1 increases giving a forward bias to the transistor, and TR$_2$ begins to conduct. When TR$_2$ conducts, it takes current away from TR$_1$ thus giving it protection. A trip switch may be incorporated to disconnect the power supply if load current increases above its safe value.

Series Thyristor Stabiliser

SWITCHING MODE POWER SUPPLY Series stabilised power units using a thyristor as the series element are known as switching mode power units. Fig.19.23 shows a block diagram using an SCR as the switching element. The SCR is triggered at regular intervals. The timing of the trigger pulse which is regulated by both the a.c. input and the d.c. output determines the d.c. level at the output (Fig.19.24). For example, if the d.c. output is high, then the trigger pulse is delayed as shown in Fig. 19.24b and vice versa as shown in 19.24c, thus compensating for these variations. Switching mode power supplies are very popular due to their simplicity, reliability, and low cost.

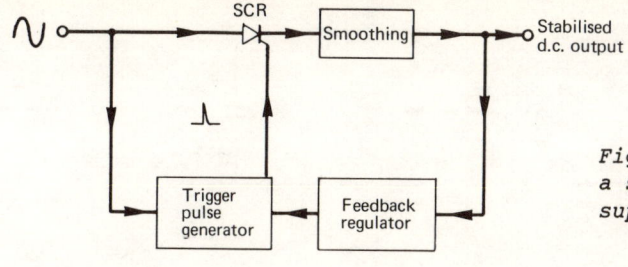

Fig.19.23 *Block diagram for a switching mode power supply*

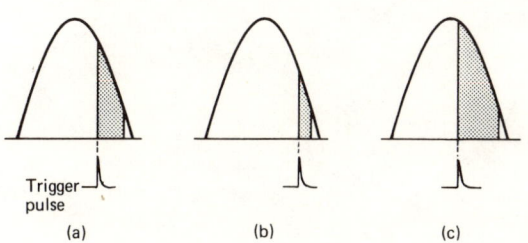

Fig.19.24 *Trigger pulse for the switching mode supply in Fig.19.23*
(a) with d.c. output normal
(b) with d.c. output high
(c) with d.c. output low

The Voltage Multiplier

In order to produce very high voltage, such as that for EHT (extra high tension) for TV receivers, voltage multipliers are used.

Fig.19.25 shows a voltage doubler. It consists of two half-wave rectifiers D_1-C_1 and D_2-C_2 connected in series. C_1 is charged by the positive half cycle (charging current i_1) and C_2 by the negative half cycle (charging current i_2) with the polarities shown. The total output voltage is therefore the sum of V_1 and V_2. If the load current is very small, both C_1 and C_2 will charge up to the peak of the input giving an output of $2V_p$.

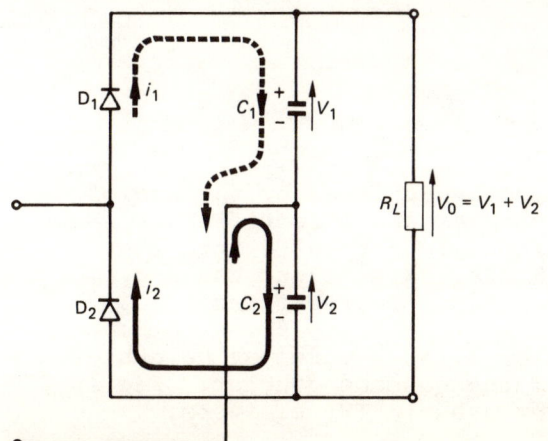

Fig.19.25 *Voltage doubler using two half-wave rectifiers D_1-C_1 and D_2-C_2 in series*

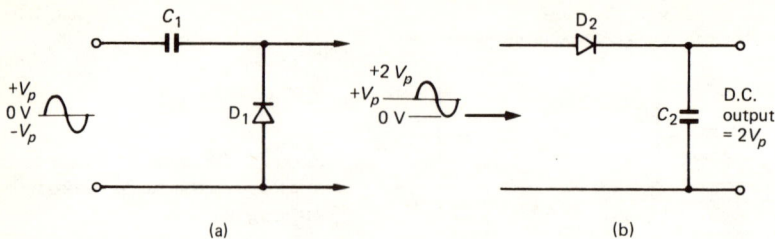

(a)   (b)

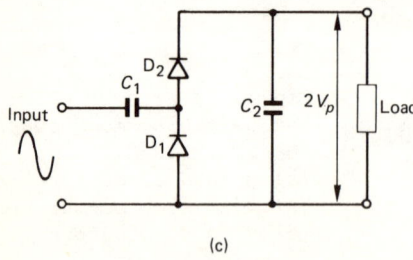

(c)

Fig.19.26 Cascade voltage doubler using a clamper C_1-D_1 and a
        rectifier D_2-C_2

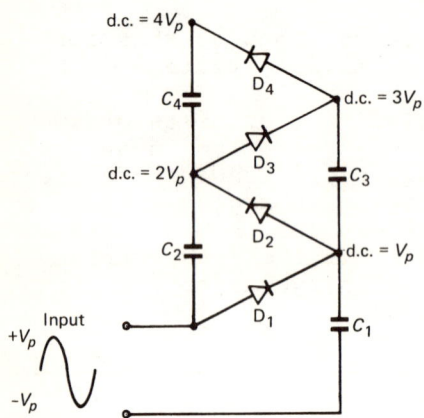

Fig.19.27 Voltage multiplier commonly used in TV receivers

Voltage doubling may also be achieved by the use of a cascade doubler using a clamper and a rectifier as shown in Fig.19.26. The a.c. input is first fed into the clamper C_1-D_1 shown in 19.26a. Its output has a d.c. level of V_p. This is then fed into rectifier D_2-C_2 charging C_2 to $2V_p$.

Voltage multiplication by 3, 4 or more may be achieved by using the principle of the cascade doubler. Fig.19.27 shows a typical multiplier used in TV receivers.

20 Amplifiers

Amplifier Classification

The operating conditions of amplifiers may be classified
into three categories.

CLASS A   The transistor or valve amplifiers so far considered have been biased in such a way as to operate on the
linear part of their characteristics.   The Q-point was
chosen in the middle to give a maximum undistorted output.
These amplifiers are said to operate under class A condition.   Fig.20.1 shows a transistor transfer characteristics.   Point A represents class A biasing.   The input
signal is small enough to keep the b-e junction forward
biased.   Hence the transistor conducts for the complete
cycle, i.e. for 360º.

   Class A amplifiers have the advantage of producing an
undistorted output, hence their wide use as pre-amplifiers
and drivers, and for i.f. and r.f. stages.   However, they
suffer from low efficiency (about 30%) due to the fact
that the transistor conducts and hence dissipates power
regardless of the presence or the absence of a signal
input.

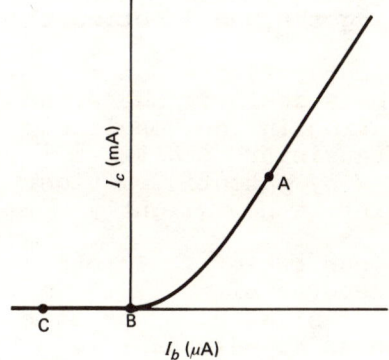

Fig.20.1 *Points A, B and C on the transfer characteristics
         representing the operating points for classes A, B and C
         respectively*

CLASS B   Here the transistor (or valve) is biased at cut
off.   Point B in Fig.20.1 shows class B biasing for a
transistor.   The transistor conducts for one half cycle
only (i.e. 180º) as shown in Fig.20.2.   Higher efficiency
is possible under class B (between 50 to 60%) since the
transistor dissipation is limited only to one half of the
input cycle.   Class B amplifiers are used in push-pull
power stages.

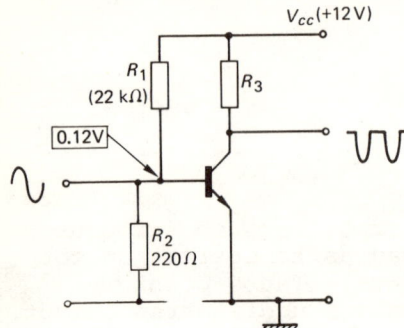

Fig.20.2 Class B amplifier: R_1-R_2 provides a small forward bias V_{be} (0.12 V) to avoid non-linearity

CLASS C The transistor (or valve) is now biased beyond cut off (point C in Fig.20.1). The transistor conducts for less than half a cycle (i.e. less than 180°). The output is therefore in the form of a pulsating waveform as shown in Fig.20.3. It has very high efficiency of between 65 to 85%. Class C amplifiers are used in oscillators and frequency changers.

Biasing

Class A biasing is provided by a potential divider as explained in previous chapters giving the b-e junction the necessary forward bias.

Class B amplifiers are biased at cut off, i.e. at a base-emitter voltage of zero. No bias chain is therefore necessary. However, to avoid operating in the non-linear part of the characteristics, the transistor is given a small forward bias of 0.1 V to 0.2 V by potential divider R_1-R_2 shown in Fig.20.2. The Q-point is now slightly above cut-off.

Class C amplifiers are biased beyond cut-off. In other words the b-e junction is given a reverse bias. This is achieved by using the input signal to bias the transistor. Hence class C biasing is also known as *signal biasing*. Two methods may be used to achieve class C biasing.

The most effective method is that shown in Fig.20.3a. Without a signal input the base is at zero potential. As shown in Fig.20.3b, the b-e junction together with C_1 and R_1 act as a clamper for incoming signals, reproducing the input waveform together with a negative d.c. of approximately peak voltage $-v_p$ as shown. The base is therefore given a reverse bias approximately equal to the peak voltage of the input. The amount of the reverse bias (or the degree of class C biasing) may be reduced by reducing the time constant C_1R_1 (usually by choosing a smaller value of R_1).

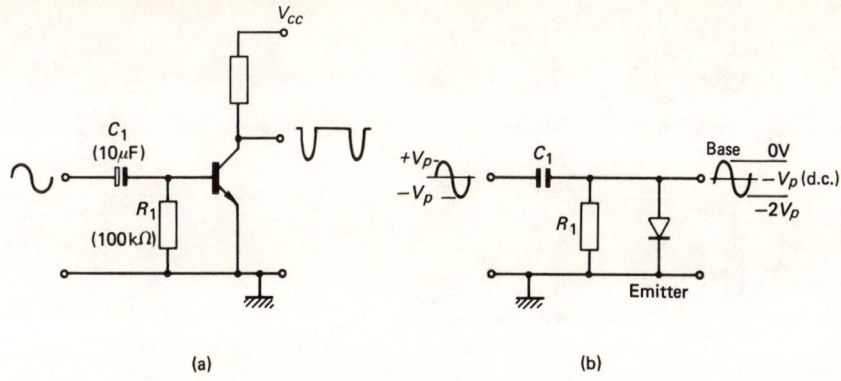

Fig.20.3 Class C signal biasing (i)
        (a) Circuit, (b) Clamping effect of C_1-R_1 and b-e junction

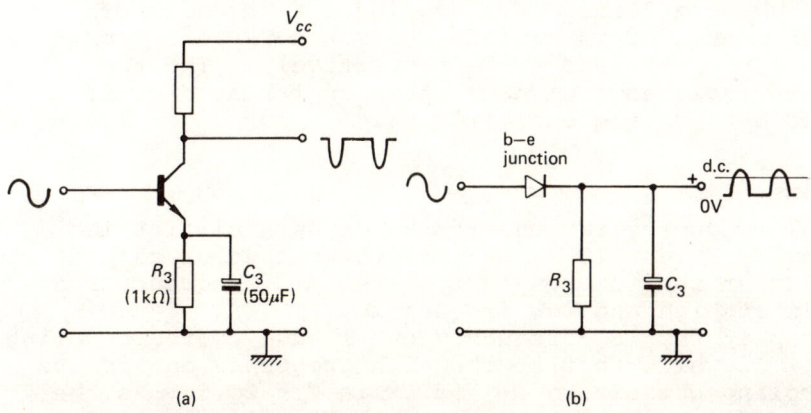

Fig.20.4 Class C signal biasing (ii)
        (a) Circuit, (b) Rectifying effect of b-e junction R_3 and C_3

A second method is that shown in Fig.20.4. Here the
emitter is given a positive potential by the charge on C_3.
With the base at zero potential, a positive emitter provides a reverse bias to the b-e junction. As shown in
20.4b for the input signal the b-e junction together with
C_3 and R_3 act as a rectifier giving C_3 a positive charge.

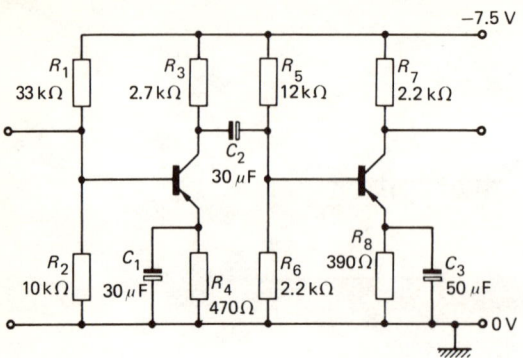

Fig.20.5 Two-stage a.f. amplifier connected in cascade

Cascade Amplifiers

Fig.20.5 shows a 2-stage $R-C$ coupled a.f. transistor amplifier connected in cascade. TR_1 and TR_2 are biased in class A by R_1-R_2 and R_5-R_6 respectively. The two stages are isolated from each other as far as d.c. is concerned by coupling capacitor C_2.

Bandwidth

A typical frequency response curve of an amplifier is shown in Fig.20.6. It shows that the output or gain of the amplifier remains constant at the mid-frequency range but falls at high and low frequencies.

The drop at the low-frequency end is due to the coupling capacitor C_2 while that at the high-frequency end is due to decoupling capacitors C_1 and C_3 in Fig.20.5. As the frequency falls, the reactance of the coupling capacitor increases, reducing the amplitude of the signal fed into TR_2. As the frequency increases, the reactance of the decoupling capacitors begin to decrease reducing the decoupling effect which introduces negative feedback, thereby reducing the output.

The fall at the high-frequency end is also due to what is known as *stray* or *interelectrode capacitors* across the transistor. Stray capacitors exist between the various electrodes of the transistor. They have the effect of reducing the gain of the amplifier at high frequencies. Every transistor has what is known as a cut-off frequency at which its current gain is too low for amplification to take place. This is an important restriction on the use of a transistor at high frequencies.

The bandwidth of the amplifier is given by the 3 dB points where the output voltage falls to 70% of its maximum value (or where the output power is half its maximum value).

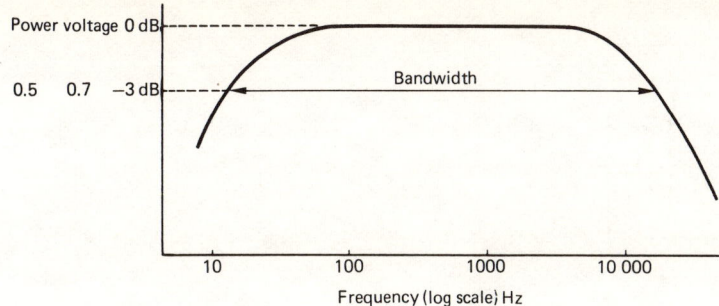

Fig.20.6 Typical a.f. frequency response

I.F. Amplifiers

The i.f. amplifier is a voltage amplifier using a tuned
circuit as the load. It operates at intermediate frequencies such as 470 kHz for a.m. radio, 10.7 MHz for f.m.
radio, and 39.5 MHz for a TV receiver.

Fig.20.7 shows a typical i.f. amplifier used in an a.m.
radio receiver. R_1-R_2 form a bias chain for TR_1, C_2 is
the bias decoupling capacitor, C_4 is the emitter decoupling
capacitor, and R_3 the emitter resistor for d.c. stabilisation. C_1-L_1 and C_3-L_2 are resonant circuits tuned to an
i.f. of 470 kHz. Transformer coupling is used for both
the input and output. The purpose of the tapping on the
primary of the output transformer T_2 is to improve the
selectivity of the i.f. amplifier. Without a tapped
primary the low output resistance r_O of the common emitter
transistor falls across the tuned circuit C_3-L_3 thus
damping it and reducing its selectivity. With tapping,
the primary winding forms a single winding step-up trans-

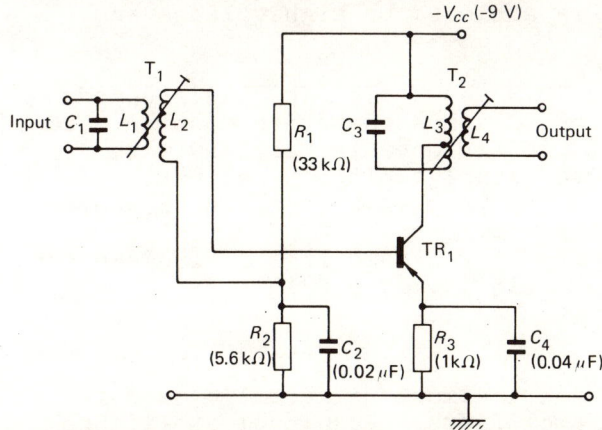

Fig.20.7 I.F. amplifier suitable for use in a.m. radio receivers

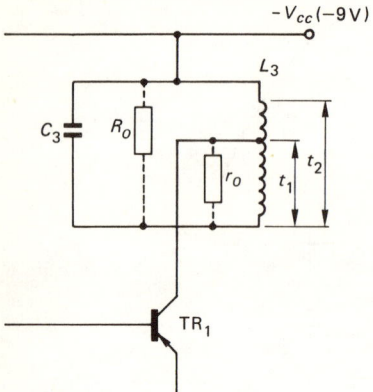

Fig.20.8 The effect of tapping a tuned circuit. L_3 is effectively an auto-transformer

former with turns t_1 as its primary and turns t_2 as the secondary as shown in Fig.20.8. The resistance shunting the tuned circuit is now the equivalent resistance to r_0 when transferred to the secondary, with a turns ratio $n = t_1/t_2$ less than one. Equivalent resistance $R_0 = r_1/n^2$ (see p.26) is larger than r_0 reducing the damping effect and improving the selectivity of the tuned circuit.

Power Amplifiers

Amplifiers so far considered are what have been referred to as voltage amplifiers, the purpose of which is to produce a large voltage swing. Where output power is required for instance to feed into a loudspeaker or an r.f. aerial, power amplifiers must be used. They have high power gain which is achieved by high voltage and current gains.

CLASS A OPERATION Fig.20.9 shows a basic single-ended transistor output stage. Class A operation must be used for undistorted output. Efficiency is very low with a large drain from the d.c. supply. It can deliver small power outputs only. It may be used in car radios where current drain is unimportant.

Higher output power may be realised from a single-ended pentode power amplifier (Fig.13.10, p.69).

Push-Pull Operation

Push-pull output stages are almost universally used in modern transistorised amplifiers. Push-pull amplifiers use two transistors operating in Class B, each one amplifying one half cycle of the input.

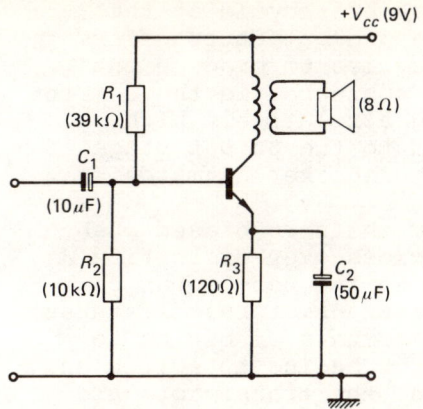

Fig.20.9 Single-ended a.f. transistor power amplifier

PUSH-PULL USING TWO SIMILAR TRANSISTORS   Fig.20.10 shows
a simplified circuit of a push-pull amplifier.   The b-e
junctions are zero biased with each transistor conducting
for alternate half cycles only.   Input transformer T_1 has
a centre tapped secondary which acts as a phase splitter.

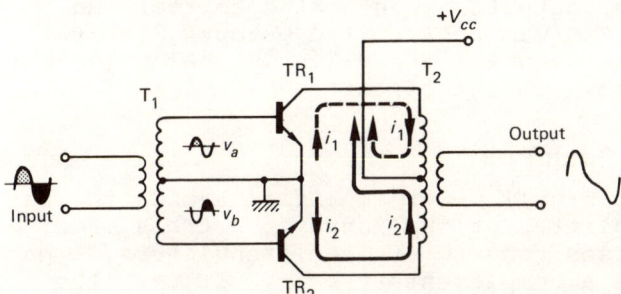

Fig.20.10 Push-pull amplifier using two identical transistors with a
transformer phase splitter

Two equal and opposite (anti-phase) signals are produced
across each half cycle of the secondary winding: v_a in
phase with the input and v_b in anti-phase to the input.
While the positive half cycle of v_a represents the positive half cycle of the input, the *positive* half cycle of
v_b represents the *negative* half cycle of the input. TR_1
and TR_2 conduct when their bases go positive with respect
to their emitters.   Thus TR_1 conducts for the positive
half cycle of v_a with current i_1 flowing as shown from
emitter to collector through the top half of output transformer T_2 towards V_{cc}.   This induces a positive half cycle
across the secondary of the T_2. TR_2 conducts for the
positive half cycle of v_b with its current i_2 flowing
upwards (opposite to i_1) through the bottom half of

transformer T_2 inducing a negative half cycle at the
secondary winding. The centre tapped output transformer
therefore combines the two half cycles to produce one
complete cycle. TR_1 and TR_2 are connected in the common
emitter configuration thus having a relatively high
output impedance. Since the load to the output stage is
usually very small, e.g. a 10Ω loudspeaker, a matching
transformer T_2 is always used.

The output waveform produced by the zero biased push-
pull amplifier suffers from the cross-over distortion
shown in Fig.20.10. This is due to the non-linear part
of the characteristics of the two transistors. It takes
place during the period when one transistor begins to
switch off and the other transistor begins to switch on.
To overcome cross-over distortion, the transistors are
given a small forward bias of between 0.1 and 0.2 V, as
shown in Fig.20.11, where R_1-R_2 provide a bias chain for
both transistors. The non-linearity of the two transistors cancel each other producing an undistorted output.

Transistor Phase Splitters

Fig.20.12 shows a phase splitter using an npn transistor.
R_3 and R_4 have equal value to give two equal and opposite
sine waves at the two outputs taken at the emitter and
the collector. For maximum undistorted outputs, the ratio
of $R_1:R_2$ should be between 2:1 and 3:1. Typical quiescent
d.c. voltages are shown on the circuit.

The Complementary Push-Pull

The complementary push-pull power amplifier avoids the use
of either a phase splitter at the input or a transformer
at the output. It uses two symmetrical transistors, a pnp
and an npn, known as a complementary pair. It uses the
fact that a positive going signal switches an npn transistor on, while a negative going signal switches a pnp
transistor on. Fig.20.13 shows the basic circuit for the
complementary push-pull. TR_1 and TR_2 are biased in class
B (i.e. at cut-off). Two d.c. supplies are used: $+V_{cc}$
and $-V_{cc}$. The positive half cycle of the input switches
TR_1 on and TR_2 off. TR_1 current i_1 flows towards $+V_{cc}$,
upwards through load resistor R as shown. For the negative half cycle, TR_2 conducts with its current i_2 flowing
away from $-V_{cc}$ towards chassis downwards through load R
in opposite direction to i_1. A whole sine wave will thus
appear across the load, corresponding to both halves of
the input cycle. It should be noted that since the output
is taken at the emitters of the transistors, TR_1 and TR_2
are therefore common collector or emitter follower amplifiers.

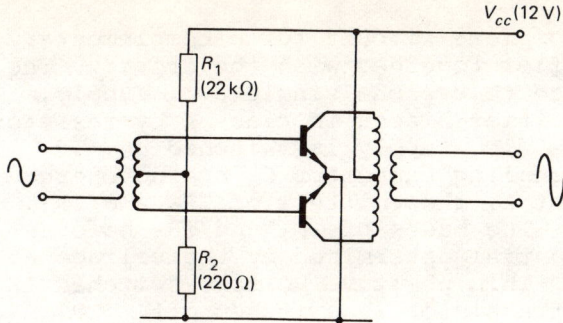

Fig.20.11 R_1-R_2 bias chain providing small forward bias to remove
cross-over distortion

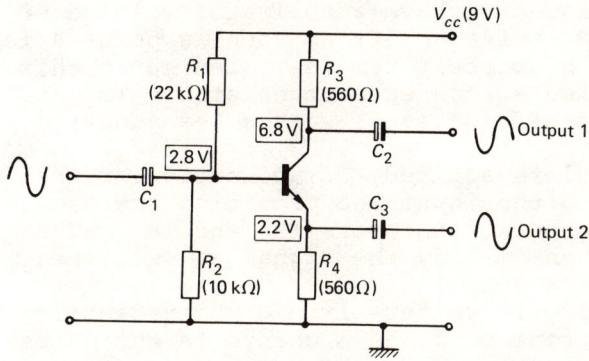

Fig.20.12 Transistor phase splitter

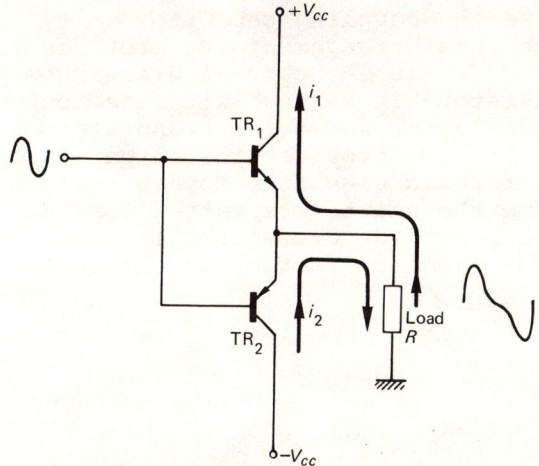

Fig.20.13 Basic circuit for push-pull complementary amplifier

Fig.20.14 shows a complete circuit for a complementary push-pull power amplifier together with the driver. The circuit is now modified to use one single d.c. supply. TR_1 is the driver amplifier biased in class A by resistor chain R_1-R_2. When the d.c. supply is switched on, TR_1 conducts normally. Coupling capacitor C_3 has no charge across it. Hence point A, the emitters of TR_2 and TR_3, is at zero potential. The bases of TR_1 and TR_2 however are at a positive potential determined by the voltage at the collector of TR_1. This positive voltage switches TR_2 on. TR_3 being a pnp transistor is switched off. TR_2 conducts as shown with its current i_2 charging C_3. As C_3 charges up, the voltage at point A rises. This continues until TR_2 switches off which takes place as soon as its emitter voltage (point A) is the same potential as its base. If TR_1 is designed to have a collector voltage of $\frac{1}{2}V_{cc}$ (4.5 V), then TR_2 switches off as soon as point A is at $\frac{1}{2}V_{cc}$ potential. The circuit remains balanced in this way with a $\frac{1}{2}V_{cc}$ applied across each transistor. TR_2 and TR_3 are biased at cut-off (class B) with a b-e junction bias of zero volts.

When an input signal is applied, TR_1 conducts for the whole cycle amplifying the input and providing the necessary drive for the output transistors TR_2 and TR_3. The complementary pair then amplify the signal as described for the basic circuit above.

The circuit in Fig.20.14 suffers from d.c. instability. Any change in TR_1 current produces a change in the quiescent condition of the output pair which may result in distortion. To avoid this, negative d.c. feedback is used to bias TR_1 as shown in Fig.20.15. The d.c. voltage at point A ($\frac{1}{2}V_{cc}$) is fed back to the base of TR_1 via resistor R_F. In this circuit the loudspeaker load is connected to the d.c. supply with C_3 as the coupling capacitor. It should be noted that with this arrangement TR_3 provides the charging current for C_3 while TR_2 current discharges it. In general the transistor "in series" with the coupling capacitor charges the latter while the transistor "in parallel" discharges it. R_4 provides the output transistors with a small forward bias to remove cross-over distortion. R_6 and R_7 are the emitter resistors for TR_2 and TR_3 respectively. They provide d.c. stability as well as a small signal negative feedback to improve the bandwidth of the amplifier.

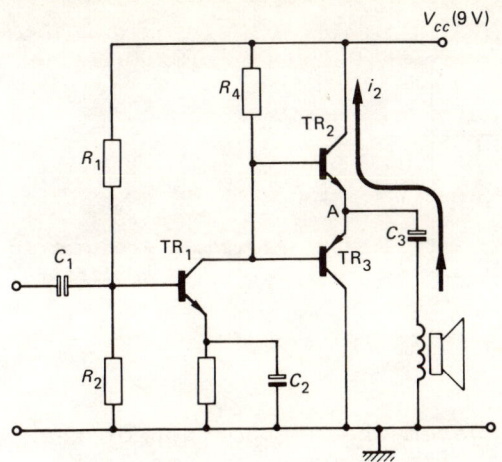

Fig.20.14 Complementary push-pull power amplifier using independent biasing chain for driver TR_1

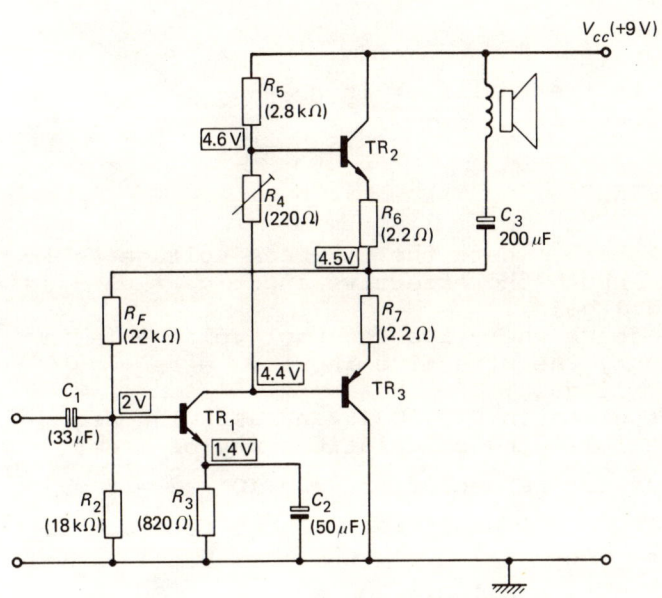

Fig.20.15 A typical complementary push-pull power amplifier. TR_1 is biased via feedback resistor R_F

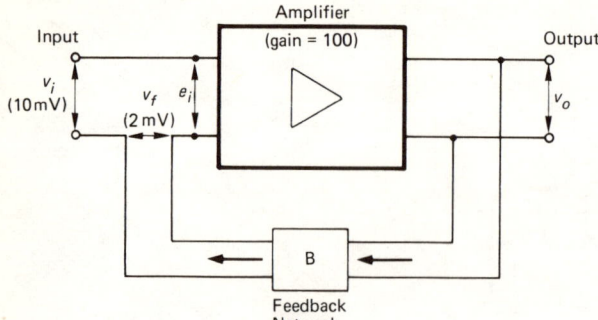

Fig.20.16 Feedback in
amplifiers

Feedback in Amplifiers

Fig.20.16 shows a feedback system where a portion of the
output voltage is fed back to the input. Voltage v_f is
the fed-back voltage which is added to the overall input
voltage v_i to produce the effective input voltage e_i. The
feedback network B feeds all or a portion β of the output
back to the input. If the output voltage is v_o then the
fed-back voltage is

$$v_f = \beta v_o$$

The overall input to the amplifier is $v_i = e_i + v_f$. Hence
the gain with feedback is

$$G_f = \frac{v_o}{v_i} = \frac{v_o}{e_i + \beta v_o}$$

For negative feedback, where the fed-back voltage is in
anti-phase to the input, the effective input $e_i = v_i - v_f$
giving a low overall gain.

For positive feedback where the fed-back voltage is in
phase with the input, the effective input $e_i = v_i + v_f$
giving a high overall gain.

Using the values given in Fig.20.16 and assuming negative feedback the following calculations may be made:

Effective input to the amplifier e_i = 10 - 2 = 8 mV

Given that the gain of the amplifier without feedback is
G = 100, then

Output voltage v_o = 8 × 100 = 800 mV

Hence the overall gain (with feedback) is

$$G_f = \frac{v_o}{v_i} = \frac{800}{10} = 80$$

The percentage feedback β is

$$\beta = \frac{v_f}{v_o} = \frac{2}{800} = \frac{1}{400} = 0.0025 \text{ or } 0.25\%$$

TABLE 20.1   Comparison between positive and negative feedback

POSITIVE FEEDBACK	NEGATIVE FEEDBACK
(1) High gain	(1) Low gain
(2) Narrow bandwidth	(2) Wide bandwidth
(3) Peaking frequency response	(3) Flat frequency response
(4) Low input impedance	(4) High input impedance
(5) High output impedance	(5) Low output impedance
(6) Introduces instability for both a.c. (hence possible oscillation) and for d.c. (hence unstable quiescent conditions)	(6) Improves stability for both a.c. and d.c.
(7) Used in oscillators	(7) Used frequently to improve stability and widen bandwidth

Feedback may also be divided into current and voltage feedback. *Current feedback* is where the fed-back voltage is proportional to the output current, e.g. that due to

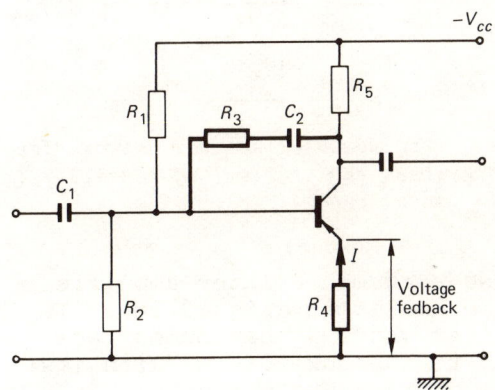

Fig.20.17 Common emitter transistor amplifier R_4 providing current feedback and C_2-R_3 providing voltage feedback

R_4 in Fig.20.17. Where the fed-back voltage is proportional to the output voltage, the feedback is known as *voltage feedback*, e.g. that due to C_2-R_3 in Fig.20.17.

Radio Frequency (r.f.) Amplifiers

At radio frequencies, e.g. v.h.f. and u.h.f., the effect of the inter-electrode capacitors, especially that between collector and base (anode and grid), becomes very noticeable. To overcome this, pentodes are used in the case of thermionic valves. In the case of transistors, the common base configuration is used. However this configuration presents a low input impedance which loads the tuner feeding the amplifier. To rectify this two methods may be used. The first is to use a common emitter amplifier with a neutralizing circuit. Such circuits cancel out or neutralize the negative feedback effect of the c-b capacitance by introducing another feedback loop in opposition to it.

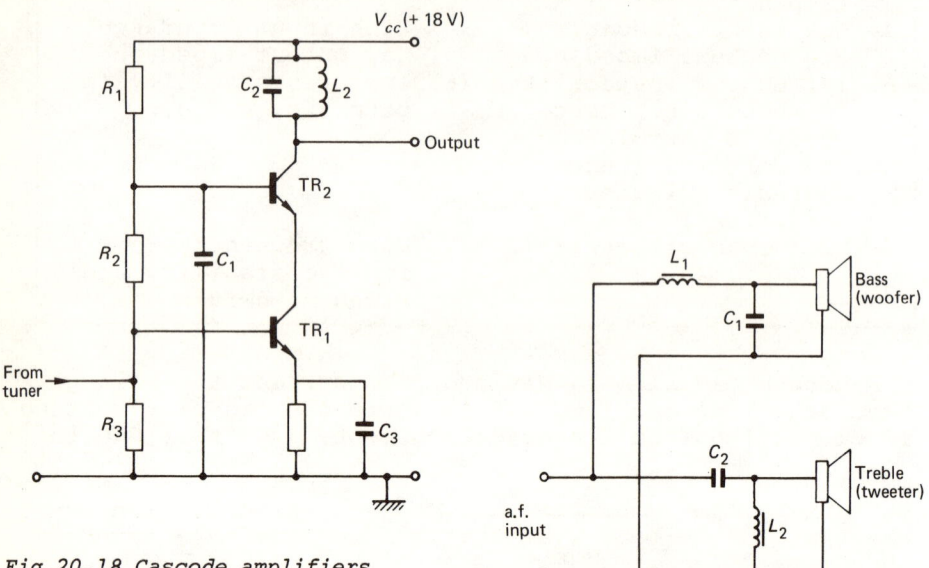

Fig.20.18 Cascode amplifiers

Fig.20.20 Cross-over network feeding bass and treble loudspeakers. (See next page)

The second method is to use a common emitter amplifier in cascode with a common base amplifier (Fig.20.18). TR_1 is the common emitter amplifier and TR_2 the common base amplifier. The input is fed to the base of TR_1 with its emitter decoupled to chassis by C_3. The output of TR_1 collector is fed to the emitter of TR_2 which has its base decoupled to chassis by C_1. The bias for TR_1 and TR_2 is provided by resistor chain R_1-R_2-R_3.

Hi-Fi

Hi-fi stands for high fidelity, a term applied to sound-producing equipment that gives a realistic re-creation of the original sound - in other words, high-quality reproduction. Hi-fi systems should therefore have a wide a.f. bandwidth 40 Hz - 16 kHz, contain low noise and low distortion.

Tone Control

The function of a tone control circuit is to extend or reduce (i.e. reshape) the frequency response curve of the amplifier with a bass control affecting the low-frequency end and a treble control affecting the high-frequency end. Tone control circuits may vary from a simple capacitor in

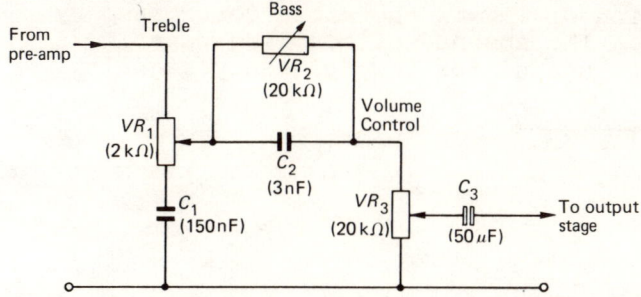

Fig.20.19 Tone control circuit

series with a resistor to a very complex arrangement employing feedback. Fig.20.19 shows a useful tone control circuit with a bass control for the low-frequency end and treble control for the high-frequency end. VR_1-C_1 offers a shunt impedance to the incoming a.f. signal. Since the reactance of C_1 is low at high frequencies, the shunt impedance attenuates high frequencies by an amount determined by the setting of VR_1. VR_2-C_2 offers a series impedance to the signal. C_2 has a high reactance at low frequencies attenuating these frequencies by an amount determined by the setting of VR_2.

Loudspeakers

A loudspeaker is a transducer that converts electrical energy into acoustic or sound energy. One of the factors affecting the choice of a loudspeaker is its frequency response, i.e. the range of frequencies it can reproduce. If the frequency range of the loudspeaker is not wide enough, then two speakers, one with good bass response and another with good treble response, may be used. Fig. 20.20 shows one possible arrangement using a cross-over network to divide the incoming frequencies range. It consists of a low pass filter L_1-C_1 feeding the bass loudspeaker (known as a woofer) and a high pass filter L_2-C_2 feeding the treble speaker (known as a tweeter).

Other factors affecting the choice of the loudspeaker are output power, efficiency, and impedance for matching purposes.

21 Other Amplifiers

D.C. Amplifiers

In direct current or d.c. amplification, direct coupling is used as shown in Fig.21.1. The base voltage of TR_2 is directly coupled to the collector of TR_1. The quiescent (no signal) condition of TR_2 is therefore determined by the previous stage. The absence of coupling capacitors make d.c. amplifiers useful for low frequency signals.

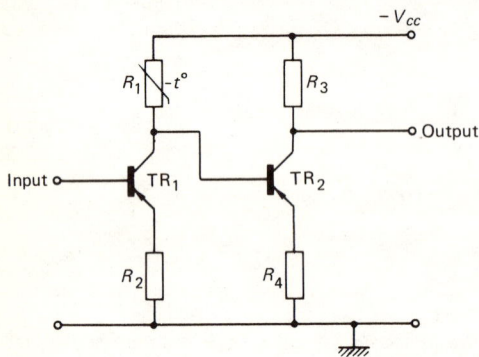

Fig.21.1 Direct coupled (d.c.) amplifiers

D.C. amplifiers suffer from what is known as DRIFT, which refers to a shift in the quiescent point of the amplifier caused by changes in temperature. To overcome the effect of drift, thermistors or other temperature sensitive elements may be used as shown in Fig.21.1.
Another method is to use the emitter coupled configuration, an example of which is shown in Fig.21.2. The effect of temperature change is greatly reduced since the drift in one stage tends to be cancelled by the other stage, leaving the output free from the effect of drift. The emitter coupled amplifier is used for other purposes such as a differential amplifier.

Differential Amplifiers

The differential amplifier has two inputs A and B, and an output which is a measure of the difference of the two inputs. Fig.21.2 shows a simple differential amplifier circuit using an emitter coupled pair of transistors. The output is taken between the two collectors.
Referring to the circuit in Fig.21.2, TR_1 and TR_2 are matched transistors and the circuit is symmetrical. Without an input the two transistors conduct equally. The current I_e through the common emitter resistor R_5 is shared equally by TR_1 and TR_2. Hence the voltages at the collectors are equal giving a zero output. A positive

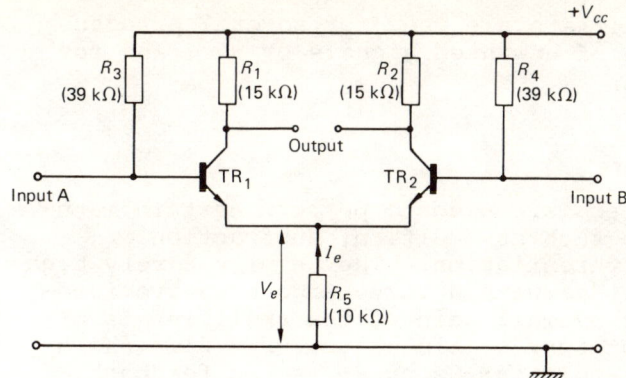

Fig.21.2 *Emitter coupled pair used as a differential amplifier*

input at A increases the current through TR_1, I_e increases, voltage V_e across R_5 increases (TR_2 emitter going towards its base), reducing the current through TR_2. Hence while TR_1 collector voltage goes down, that at TR_2 collector goes up producing an output. A negative input at A produces the opposite effect on the voltages at the two collectors, resulting in an opposite output. Since the circuit is symmetrical, an input at B produces similar results. When an input is fed into both A and B, the output will be a measure of the difference of the two inputs. Although only d.c. inputs have been considered, the circuit functions equally as well for a.c. inputs.

For good amplification R_5 should have as high a value as possible. A transistor is sometimes used in place of the common emitter resistor as shown in Fig.21.3. Transistor TR_3 has a constant bias provided by $-V_{BB}$ and resistor chain R_1-R_2, thereby producing a constant current I_3. TR_3 thus represents a constant current source and hence a very high resistor.

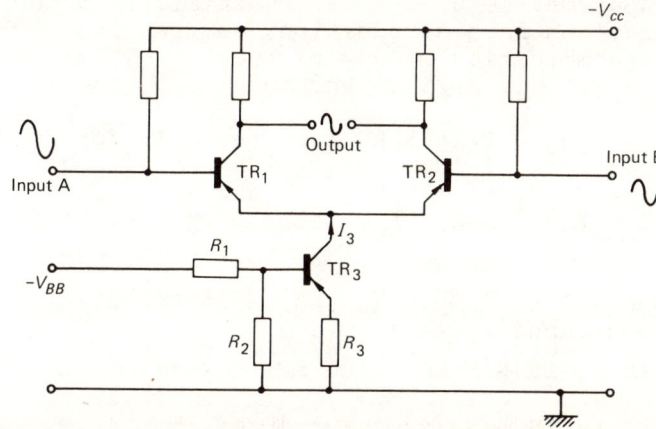

Fig.21.3 *Differential amplifier employing an emitter coupled pair with TR_3 as a constant current device*

Differential amplifiers are used for several purposes such as elimination of unwanted signals or hum, and for power stabilisation.

Operational Amplifiers

Operational amplifiers are used to perform certain mathematical operations such as addition, subtraction, integration and differentiation. They employ a very high gain amplifier together with a very large negative feedback. This way the overall gain of the amplifier is reduced to a low but very stable value. Fig.21.4 shows a simple operation amplifier using R_F as the feedback resistor. It can be shown that if the feedback is very large, the overall gain of the amplifier is very stable and is equal to the ratio R_F/R_1. For example, if R_F = 20 kΩ and R_1 = 2 kΩ then the gain 20 k/2 k = 10. The operation amplifier may now be used as a ×10 multiplier. Thus for an input voltage V_{in} = 1 V, the output V_O = 10 × V_{in} = 10 × 1 = 10 V.

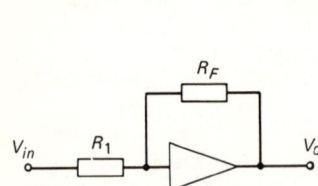

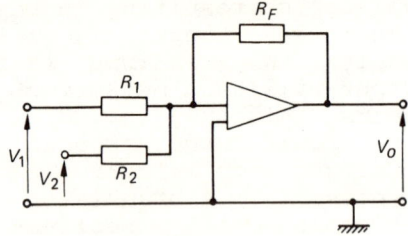

Fig.21.4 Operational amplifier with a single input

Fig.21.5 Operational amplifier with two inputs

More than one input may be used as shown in Fig.21.5 to provide other operations such as addition, subtraction and division, or a combination of any of them.

For input V_1, the gain is R_F/R_1. For input V_2, the gain = R_F/R_2.

For example if R_F = R_1 = R_2 = 5 kΩ then the gain for both inputs = 5 kΩ/5 kΩ = 1. Given V_1 = 1 V and V_2 = 2 V we get an output due to V_1 of 1 × 1 = 1 V and an output due to V_2 of 1 × 2 = 2 V. Hence the total output V_O is 1 + 2 = 3 V.

EXAMPLE In the operational amplifier shown in Fig.21.6 input V_1 = 20 mV and input V_2 = -10 mV.

The output due to V_1 is $\frac{5}{1}$ × 20 = 100 mV.

The output due to V_2 is $\frac{5}{5}$ × (-10) = - 10 mV.

Hence the total output V_O = 100 - 10 = 90 mV.

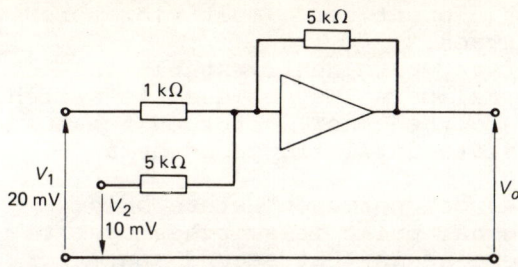

V_1
20 mV

1 kΩ

5 kΩ

V_2
10 mV

V_o

Fig.21.6

Wideband Amplifiers

A wideband amplifier is one that is capable of providing amplification over a wide range of frequencies extending from a few hertz to several megahertz. Where a coupling capacitor is used, the lower frequency limit of the useful range is determined by the coupling capacitor. The upper frequency limit is determined by the high frequency response of the transistor and is known as its upper cut-off frequency.

Given the upper-cut frequency it is possible to extend the bandwidth by using compensating networks or neutralising circuits to neutralise the high frequency effects of the transistor.

22 Multivibrators

Multivibrators consist of two transistors (or valves) arranged so that one transistor is always fully conducting and the other is cut off. Hence a multivibrator has two distinct stable states: TR_1-on/TR_2-off and TR_1-off/TR_2-on. Fig.22.1 shows a basic diagram for a multivibrator circuit where z_1 and z_2 are two coupling elements providing positive feedback.

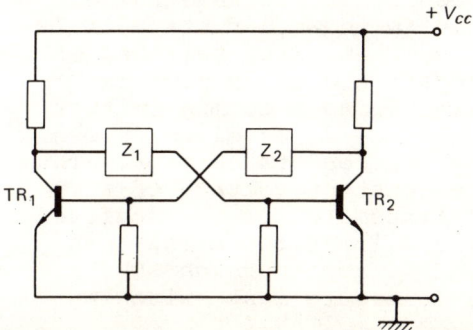

Fig.22.1 Basic schematic diagram for a multivibrator

Depending on the feedback element used, multivibrators
may be divided into three types.
(1) the BISTABLE which can stay permanently stable in
     either state. An external pulse may be used to switch
     it over from one state to the other. It then remains
     stable in that second state until triggered by a
     second pulse and so on.
(2) The MONOSTABLE which has one permanent state only.
     If triggered by an external pulse it switches over to
     the other state. It remains in that second state for
     a duration determined by the time constant of the
     feedback elements, and returns back to the original
     state by itself.
(3) The ASTABLE which is a free running oscillator having
     no permanent state. It continually changes from one
     state to the other and back again and so on.

The Bistable Circuit

Referring to Fig.22.2, when the d.c. supply is switched
on, because of component tolerances, etc. one transistor
will tend to conduct more than the other. However small
the difference in transistor currents it is enough to
define the state in which the bistable will settle in.
Supposing that TR_2 begins to conduct more than TR_1.
TR_2 collector voltage will drop causing a drop in TR_1 base
voltage, TR_1 current decreases, its collector voltage
increases causing TR_2 base to go up away from the emitter.
TR_2 current increases further and so on until TR_2 is fully
on (saturates) and TR_1 is fully off. In this state TR_1
being at cut-off has a collector voltage of $+V_{cc}$ (10 V),
while TR_2 being at saturation has a collector voltage of
zero volts. The voltage at the base of TR_1 is determined
by resistor chain R_3-R_5. As shown in Fig.22.2b, the base
is kept at a negative potential by negative supply $-V_{BB}$
keeping TR_1 off. The voltage at the base of TR_2 is
determined by resistor chain R_2-R_6. As shown in 22.2c
TR_2 base has a positive voltage, giving TR_2 the necessary
forward bias. The bistable will thus remain in this
state permanently if undisturbed by an external pulse.
It may remain equally as stable in the other state if TR_1
started conducting more than TR_2 when first switched on.
To avoid using a separate negative d.c. supply, the
circuit in Fig.22.3 is used with R_6 as a common emitter
resistor. In either state a voltage V_e develops across
R_6 due to the current taken by the "on" transistor. Thus
the two emitters are kept at a constant voltage of V_e.
The reverse bias of the "off" transistor is now achieved
by ensuring that its base is at a lower potential than
its emitter. Capacitors C_2 and C_3 are known as SPEED-UP
CAPACITORS. Their purpose is to ensure fast switching
from one state to the other.

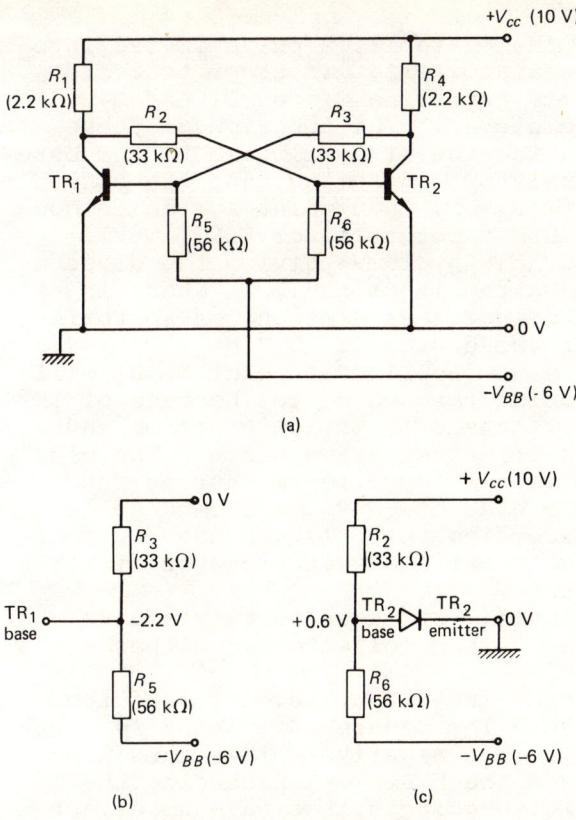

(a)

(b)   (c)

Fig.22.2 Bistable multivibrator: (a) circuit using separate d.c.
    supplies V_{cc} and - V_{BB}

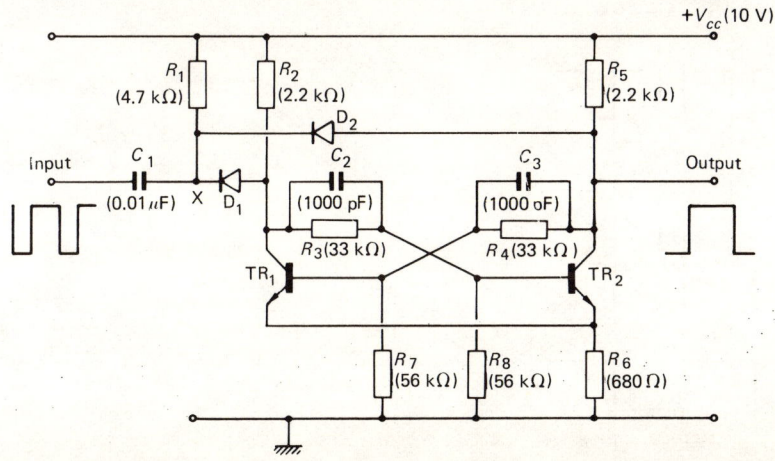

Fig.22.3 Bistable multivibrator with steering diodes D_1 and D_2

STEERING DIODES

To change the state of the bistable, a pulse is fed into it turning the "on" transistor off. In order to avoid using two separate inputs, steering diodes D_1 and D_2 shown in Fig.22.3 are employed. The function of these diodes is to steer or guide the trigger pulse to the base of the appropriate transistor. Assuming that the bistable is in the state TR_1-off/TR_2-on, then point X, the common cathodes of D_1 and D_2, has a potential of V_{cc} (10 V). The anode of D_1 is also at V_{cc} (10 V) giving the diode a bias voltage of zero. On the other hand, D_2 anode is at emitter potential (TR_2 being at saturation) of approximately 1 V giving D_2 a reverse bias of -9 V.

If a negative pulse is now applied at point X, D_1 will conduct directing the pulse through R_3 to the base of TR_2 turning it off. As TR_2 turns off, TR_1 switches on and the bistable settles in its alternative state. The bias of the steering diodes is now opposite to what it was with D_1 having a reverse bias of -9 V. A second pulse will forward bias D_2 directing the pulse through R_4 to the base of TR_1 turning it off and switching the bistable back to its original state.

As the bistable continually switches over, a square wave output may be taken at the collector of either transistor.

Fig.22.4 shows the input and output waveforms of the bistable described above. The square wave input is first differentiated by C_1-R_1. The negative spikes thus produced are used to trigger the bistable, producing the output shown. The positive-going spikes have no effect on the circuit as both diodes will be reverse biased. The output is another square wave having a frequency half that of the input. This is why the bistable is known as a divide by two ($\div 2$) device which is extensively used in counters and calculators.

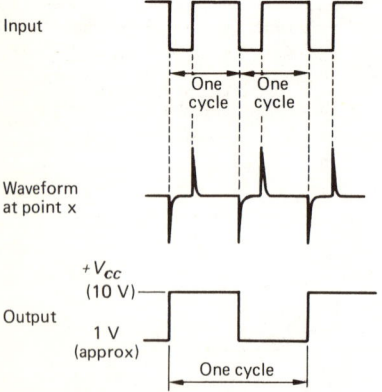

Fig.22.4 Input and output waveforms of a bistable multivibrator

The Monostable

The feedback loop of the monostable (Fig.22.5) contains one C-R network, C_2-R_2. When the circuit is first switched on TR_2, having its base taken towards V_{CC} via R_2, conducts very heavily, turning TR_1 off in the process. The negative d.c. supply $-V_{BB}$ ensures that TR_1 remains off. This is the permanent state of the circuit.

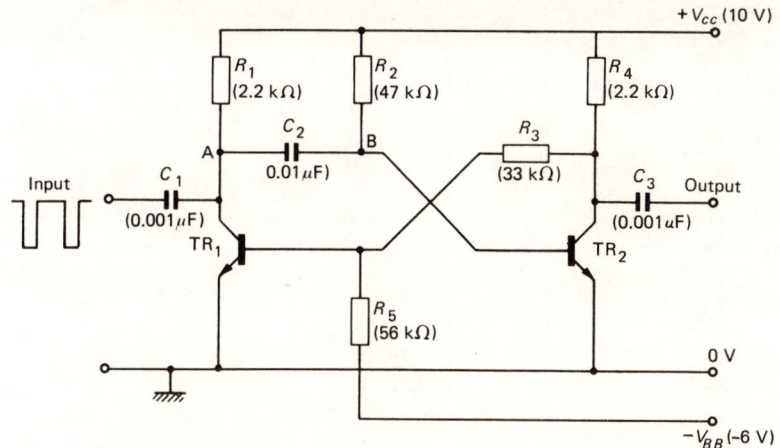

Fig.22.5 Monostable multivibrator

If a negative pulse is now applied at the input, the high frequency edge will pass through C_2 to the base of TR_2 turning it off. In a similar way to the bistable, TR_1 will saturate and TR_2 will cut off. The collector of TR_1 (point A) previously at +10 V (V_{CC}) goes down sharply to zero volts, a drop of 10 V giving capacitor C_2 a charge of -10 V. In other words TR_2 base (point B) is now at -10 V holding the transistor off. Capacitor C_2 begins to discharge through R_2 from -10 V attempting to charge up to +10 V. The negative potential at point B begins to decrease gradually at a rate determined by time constant C_2R_2. As point B crosses the zero line (Fig.22.6b), TR_2 begins to conduct turning TR_1 off, and the monostable is back to its original state, awaiting the second pulse.

The output of the circuit is a pulse waveform shown in Fig.22.6c. The pulse width is determined by the period during which TR_2 remains in the off state, itself determined by time constant C_2R_2. The circuit in Fig.22.5 for example gives a pulse duration of approx. 350 µs. This may be changed by varying the value of C_2 or R_2 or both. It should be noted that while the frequency of the output is the same as that of the input, the pulse width is different. Monostables are used to increase the width of a pulse and for time delay as will be explained in section 29.

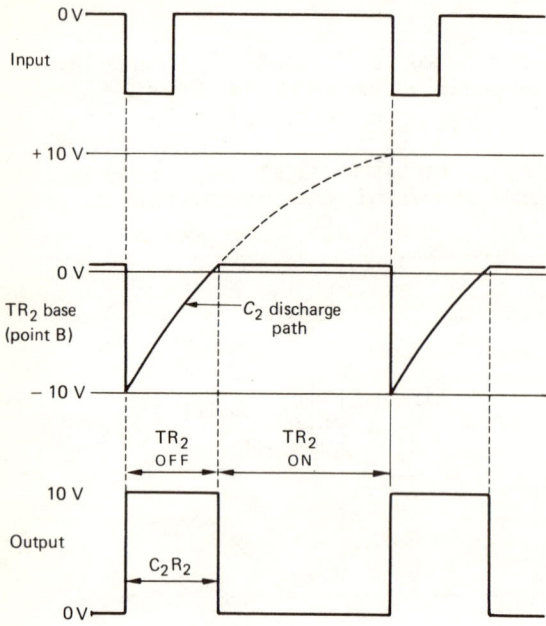

Fig.22.6 Monostable waveforms

The Astable

Fig.22.7 shows an astable multivibrator. The feedback loop contains two time constant circuits C_1R_1 and C_2R_2. When the circuit is first switched on, one transistor tends to conduct more than the other. Due to the feedback loop this results in one transistor fully on while the other is fully off. Suppose that TR_1 turns on and TR_2 off with C_1 charging up to $-V_{cc}$ keeping TR_2 off. Capacitor C_1 begins to discharge through R_1 attempting to charge up to $+V_{cc}$ as in the case of the monostable. As the potential of the junction R_1-C_1 (TR_2 base) passes through zero, TR_2 switches on turning TR_1 off. C_2 now charges up in the negative direction keeping TR_1 off. As C_2 discharges through R_2 it turns TR_1 on and so on. The circuit oscillates and produces a square wave output at either collector. The mark-and-space times are determined by the time constants of the circuit.

Fig.22.8 shows the output from the collector of each transistor. For equal mark-and-space, C_1R_1 and C_2R_2 are made equal. Note that as in the case of the monostable, the *CR* time constant feeding into the base of a transistor determines the time that the transistor remains off.

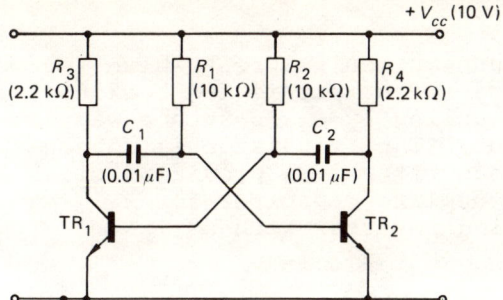

Fig.22.7 Astable multivibrator

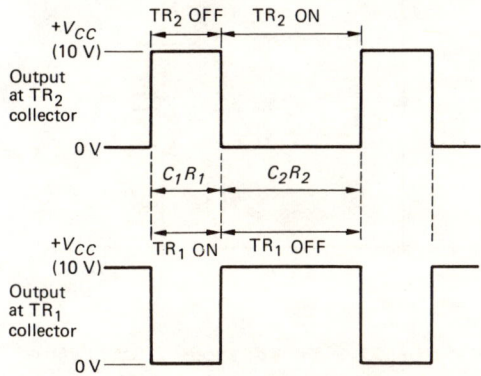

Fig.22.8 Outputs from the collector of each transistor of the astable in Fig.22.7

23 Oscillators

The oscillator is an amplifier with adequate positive feedback to produce an output without an external input. It converts d.c. (the supply to the amplifier) to a.c. signal. Two basic requirements are necessary for sustained oscillations:

(*a*) positive feedback and

(*b*) a gain round the feedback loop of more than one.

There are two types of oscillators: the sinusoidal oscillator producing sine wave outputs; and the non-sinusoidal, also known as relaxation oscillator, normally producing square waves.

TUNED OSCILLATORS   These are sinusoidal oscillators using an L-C tuned circuit in the feedback loop to determine the frequency of oscillation.

Tuned Collector Oscillators

In Fig.23.1, L_2-C_2 is the tuned circuit across which the output is developed. Part of the output is fed back to the input via transformer coupling L_2-L_1 in such a way as to be in phase with the input. The transistor is connected in the common emitter mode with R_1-R_2 providing class A biasing. C_1 is a bias decoupling capacitor for R_2. R_3 is the normal d.c. stabilising resistor with C_3 its decoupling capacitor.

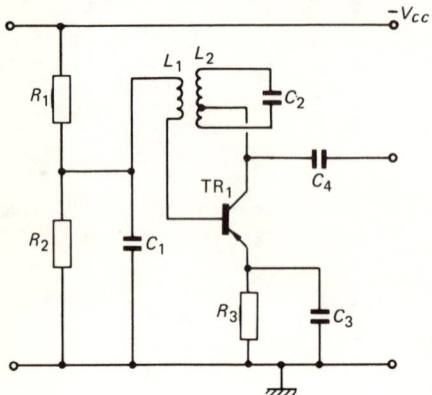

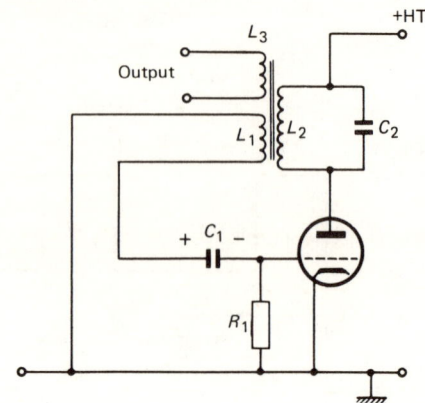

Fig.23.1 Tuned collector
oscillator

Fig.23.2 Tuned anode oscillator

Tuned Anode Oscillators

A common cathode triode is used in class C. In Fig.23.2, C_1-R_1 provides the class C bias known as the leaky grid bias. It is similar in principle to the biasing arrangement described on page 41 for a transistor. Capacitor C_1 charges up as shown giving the control grid a negative bias beyond cut-off.

Apart from its high efficiency and high power output class C biasing provides stability to both the amplitude and frequency of the output.

Tuned Base Oscillators

In Fig.23.3, C_2 is a coupling capacitor which provides class C biasing for TR$_1$. L_2-C_1 is the tuned circuit. Positive feedback is achieved via C_3 and transformer T_1.

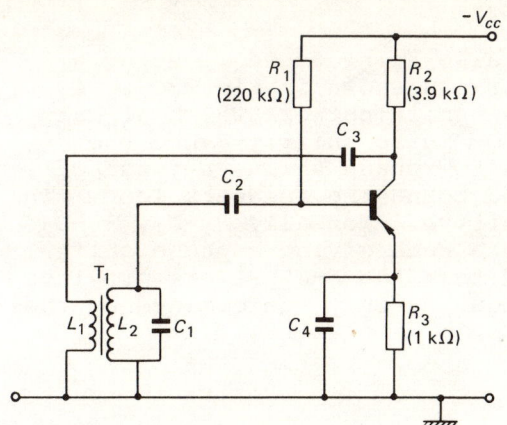

Fig.23.3 *Tuned base oscillator*

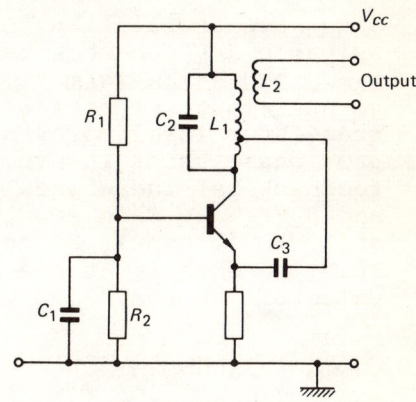

Fig.23.5 *Hartley oscillator*

Tuned Grid Oscillators

In Fig.23.4, C_2-R_2 provides class C biasing with the output taken across transformer winding L_3.

Hartley Oscillator

This oscillator (Fig.23.5) uses a split inductor L_1 to provide the necessary feedback, which is fed into the emitter. C_2-L_1 forms the tuned circuit.

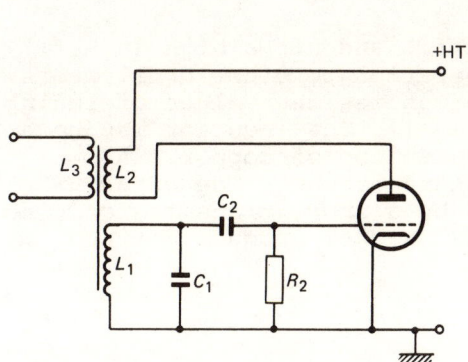

Fig.23.4 *Tuned grid oscillator*

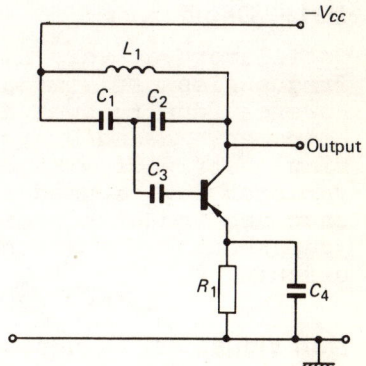

Fig.23.6 *Colpitt oscillator*

Colpitt's Oscillator

A split capacitor C_1-C_2 is used in this case as in Fig. 23.6. C_1-C_2-L_1 forms the tuned circuit, with C_3 providing class C bias.

R-C or Phase Shift Oscillators

Sine wave oscillations may also be produced by using a
suitable $R-C$ feedback network as shown in Fig.23.7. R_1-C_1,
R_2-C_2 and R_3-C_3 form a phase shift network which at one
given frequency provides 180° phase shift. Since the
transistor itself provides 180° phase shift, a total of
360° phase shift is produced round the feedback loop. The
feedback is, therefore, positive. Normally R_1-C_1, R_2-C_2
and R_3-C_3 are made equal with each giving a phase shift
of 60°. It should be noted that the $R-C$ network provides
180° phase shift at one frequency only, determined by the
value of the components.

Crystal Oscillators

One very important requirement of an oscillator is the
stability of its frequency. Changes in frequency may be
caused by changes in the values of the tuned circuit
components L and C or changes in transistor parameters
due to temperature variations. Frequency stability may
be improved by the proper choice of the circuit elements
including the transistor. For very high stability a
quartz crystal is used as shown in Fig.23.8, with the
crystal determining the frequency of oscillation. However, small variations may be achieved by placing a
variable capacitor C across the crystal. Crystal oscillators are used in colour TV receivers to produce the
4.43 MHz sub-carrier to an accuracy of a few Hz.

U.H.F. Oscillators

Oscillators at very high (v.h.f.) and ultra high (u.h.f.)
frequencies are similar in operation to other oscillators.
However, due to such high frequencies the values of tuning
components L and C are very small. The inductor may be a
single strip of wire or a simple loop of copper. A
varactor may be used for the capacitance. The capacitance and inductance displayed by a short transmission line
(see section 28) are sometimes used to produce a tuned
circuit.

Non-sinusoidal Oscillators

Also known as relaxation oscillators these produce a
square or pulse output by switching one or two transistors
on and off. The astable multivibrator is one such oscillator described in the previous section. Another type is
the blocking oscillator.

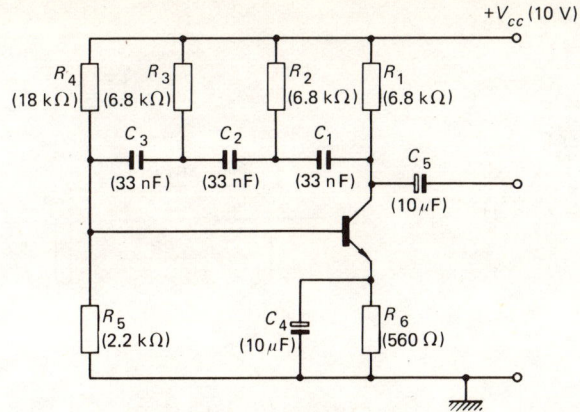

Fig.23.7 *Phase shift oscillator.* R_1-C_1, R_2-C_2, R_3-C_3 *provide 180°
phase shift*

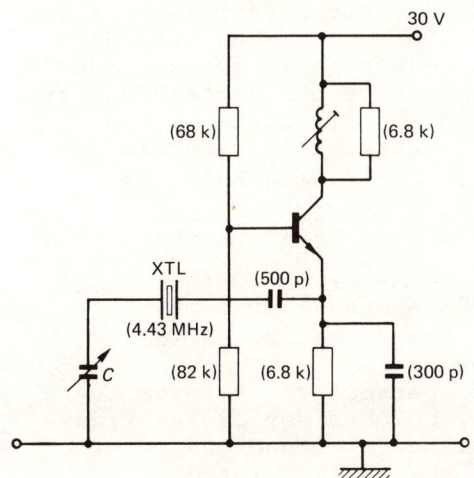

Fig.23.8 *Crystal oscillator. The circuit is used as the reference
oscillator in TCE 3000 colour TV receiver. Note that the
self-capacitance of the inductor is used for tuning.
Variable capacitor C is in fact a varactor. (By permission
of Thorn Consumer Electronics)*

Blocking Oscillators

This type uses a transformer coupled feedback from collector to base (Fig.23.9). The action of the circuit hinges
around the fact that, due to transformer coupling, a
voltage is induced into the base only when collector
current is varying, i.e. increasing or decreasing. In
the one case the feedback is positive and in the other it
is negative. When the circuit is first switched on the

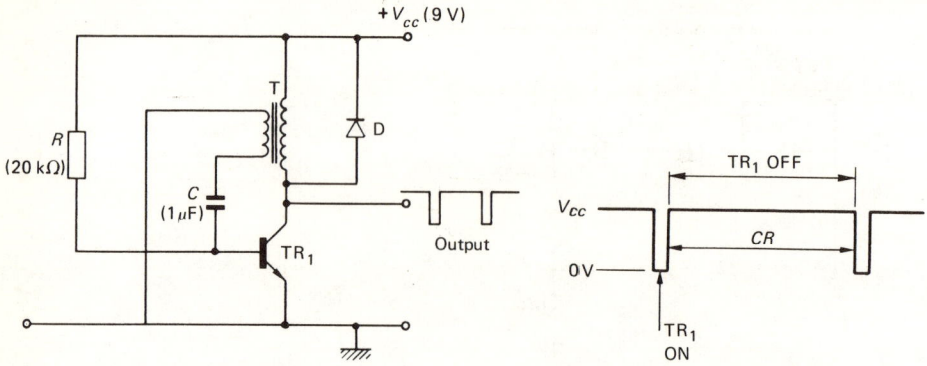

Fig.23.9 Blocking oscillator

Fig.23.10 Output of a
blocking oscillator

transistor conducts and collector current increases thereby producing a feedback voltage at the base in such a way
as to switch the transistor further on. When saturation
is reached, collector current ceases to increase, inducing
a voltage at the base which this time turns the transistor
off. The transistor is held in the off state by the
negative charge on capacitor C, until the latter is
sufficiently discharged through resistor R when the
transistor switches on again and so on.

The output from a blocking oscillator is a narrow pulse
waveform as shown in Fig.23.10. The width or mark of the
pulse is determined by the transformer parameters while
the space is determined by time constant CR. The
frequency of oscillation may, therefore, be varied by
changing the value of resistor R.

Depending on the transformer parameters, a large overshoot in collector voltage may be obtained as the transistor switches off. This overshoot voltage may be of
such a magnitude as to exceed the maximum rateable
collector voltage resulting in damage to the transistor.
To protect the transistor a diode D_1 is connected across
the primary winding of the transformer as shown. The
diode is normally reverse biased. It is forward biased
only if the collector voltage exceeds that of the d.c.
supply V_{cc}. This way the voltage at the collector cannot
exceed that of the supply.

Unijunction Oscillators

Devices having negative resistance characteristics such
as the unijunction or the tetrode readily lend themselves
for use in oscillators. Fig.23.11 shows an oscillator
using a unijunction. The unijunction is biased to operate along that part of its output characteristics where

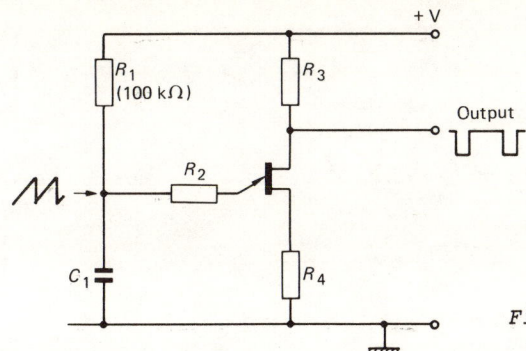

Fig.23.11 Unijunction oscillator

the output current increases as the output voltage
decreases, i.e. negative resistance. It continually
switches on and off without any feedback. The output at
base 2 (b_2) is a pulse waveform. A sawtooth output may
also be taken at the emitter. The periodic time or
frequency is determined by time constant C_1R_1.

Sawtooth Generators

Fig.23.12 shows a sawtooth generator using a pulse input.
For the period of the input cycle between A and B (Fig.
23.13) the base of the transistor is at zero voltage and

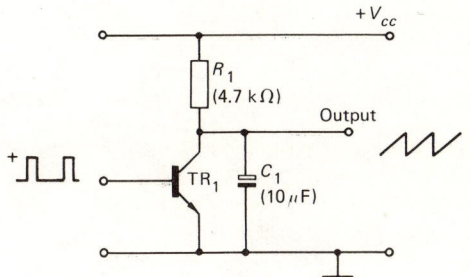

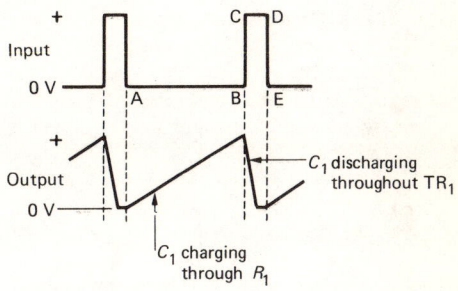

Fig.23.12 Sawtooth waveform
produced from a pulse input

Fig.23.13 Input and output waveforms
of the sawtooth generator in
Fig.23.12

the transistor is cut off. Capacitor C_1 thus steadily
charges up through resistor R_1. Before the capacitor is
fully charged, positive edge BC arrives switching TR_1 on
and discharging C_1 through the transistor very rapidly.
The capacitor remains discharged for the duration of the
pulse from C to D. Negative edge DE switches the transistor off and C_1 begins to charge up again and so on.

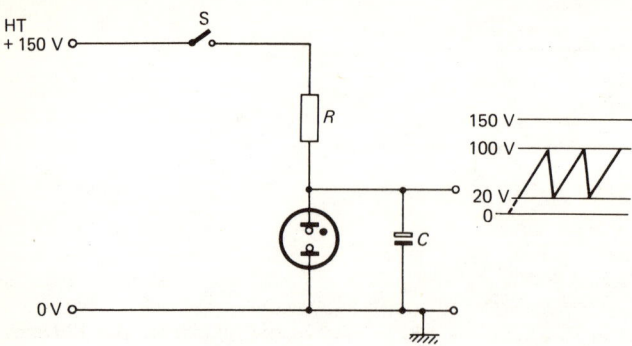

Fig.23.14 *Sawtooth generator using a neon diode*

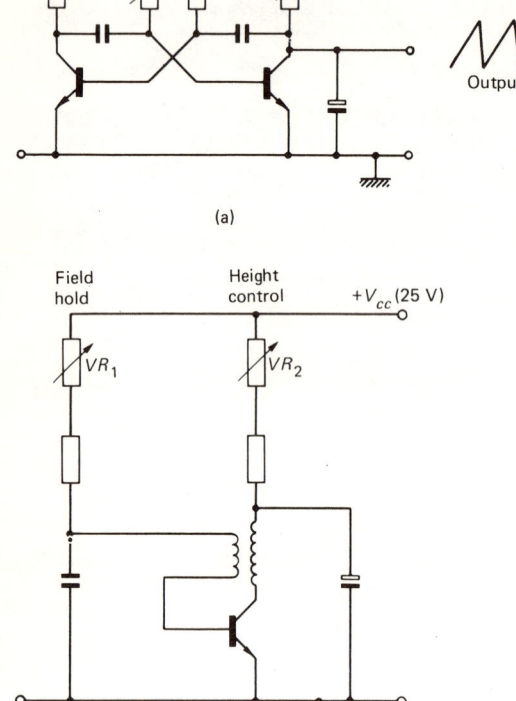

(a)

(b)

Fig.23.15 *Sawtooth generators used in the field timebase of TV*
        receivers
        (a) using an astable multivibrator
        (b) using a blocking oscillator

A sawtooth waveform may also be produced using a gas-filled diode as shown in Fig.23.14. Assuming the neon diode has a striking voltage of 100 V and an extinction voltage of 20 V, then as switch S is closed the capacitor charges up through resistor R towards HT (= 150 V). When the neon conducts at 100 V, the capacitor discharges very rapidly to 20 V. The neon stops conducting and the capacitor starts charging up again and so on. The output is a sawtooth waveform varying between 20 V and 100 V.

The same principle of charging and discharging a capacitor is used in other sawtooth generators. Fig.23.15 shows two such generators using an astable multivibrator and a blocking oscillator respectively. Such oscillators are used in TV receivers for the field timebase. VR_1 controls the frequency (field hold) and VR_2 controls the amplitude (height) of the waveform.

24 Frequency Changing

There are two methods used for frequency changing.
(1) The ADDITIVE method where two different frequencies f_1 and f_2 are added $f_1 + f_2$ or subtracted $f_1 - f_2$ to produce a third frequency.
(2) The MULTIPLICATIVE method where one frequency is multiplied by a factor like 2, 3, etc.

The Mixer-Oscillator

One circuit commonly employed for frequency changing using the additive method is the mixer-oscillator. It is used in radio and TV receivers to produce the intermediate frequency i.f. Fig.24.1 shows such a circuit suitable for an a.m. radio receiver. The r.f. carrier f_1 is selected by tuned circuit $VC_1-C_5-L_1$ and fed into the base of TR_1 via transformer coupling L_1-L_2 and coupling capacitor C_1. The output developed across L_4 is fed back via transformer coupling $L_4-L_5-L_3$ into the emitter of the transistor. The feedback is positive, producing oscillations. The

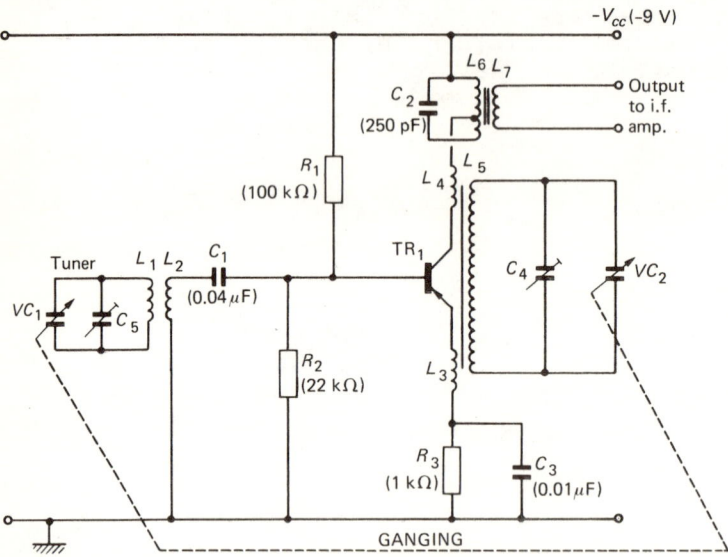

Fig.24.1 *Mixer oscillator circuit suitable for a.m. radio receivers*

frequency of oscillation f_0 is determined by the tuned circuit $L_5-C_4-VC_2$. The two frequencies f_1 and f_0 are mixed together to produce the original two frequencies as well as their sum $f_0 + f_1$ and their difference $f_0 - f_1$. The collector load C_2-L_6 is tuned to the difference $f_0 - f_1$ which is the i.f.

VC_1 and VC_2 are ganged together so that the oscillator frequency is always higher than the selected r.f. carrier by the i.f. (470 kHz for a.m. radio receivers). In this way an output having a constant frequency, i.e. the i.f., is always produced no matter what the selected carrier frequency is. The transistor for a mixer-oscillator is normally biased in class C. In the circuit shown in Fig. 24.1 class C biasing is achieved by capacitor C_1, which together with bias chain R_1-R_2 give the transistor a small reverse bias.

The Frequency Multiplier

Fig.24.2 gives the circuit. Transistor TR_1 is biased in class C by capacitor C_1. The sinusoidal input is thus distorted, producing a number of harmonics of the input frequency. C_2-L_1 is tuned to a multiple of the input frequency, e.g. the second harmonic, third harmonic, etc. Provided the gain of the amplifier is high enough at that particular harmonic, and provided the tuned circuit has a high Q-factor, a sustained sinusoidal output may be produced having a frequency which is a multiple of the input frequency.

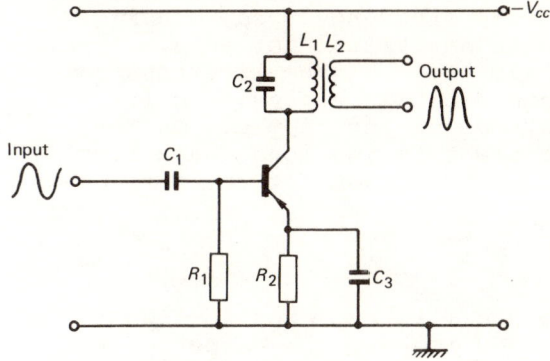

Fig.24.2 Frequency multiplier employing class C amplifier

25 Demodulation

As explained in *Principles and Systems for Radio and TV Mechanics*, in radio transmission the information or the signal is used to modulate a carrier of radio frequency. The modulated carrier is then transmitted. A receiver tuned to the carrier frequency changes the received carrier frequency to an intermediate frequency, i.f. The modulated i.f. is then amplified and detected or demodulated to extract the information signal, e.g. the a.f. for a radio receiver or the video for a TV receiver from the modulated i.f. The frequency changing is carried out by using the mixer-oscillator circuit described in section 24. Demodulation circuits are described below.

The Amplitude Modulation Detector

Amplitude modulation detector circuits are similar to a simple d.c. power unit. Fig.25.1 shows a typical circuit used in a.m. radio receivers. The modulated i.f. is first rectified by diode D_1. The diode used is normally germanium (referred to as a signal diode) because of its low forward voltage drop. Capacitor C_1 charges up to the peak of the carrier i.f. and discharges gradually through the volume control resistor VR_1 as shown in Fig.25.2b. The r.f. ripple produced is smoothed or filtered out by low pass filter R_1-C_2 to reproduce the envelope of the modulated wave as shown in 25.2c.

The length of the time constant C_1VR_1 is critical, depending on the intermediate frequency used. In Fig. 25.1 the values shown are those typical for an a.m. radio receiver with an i.f. of 470 kHz. If the time constant is too small, C_1 discharges rapidly between each i.f. peak, increasing the amplitude of the ripple. On the other hand if the time constant is too long, capacitor C_1 will discharge very slowly missing some of the i.f. peaks as shown in 25.2d.

Automatic Gain Control

Automatic gain control (a.g.c.) is employed for two reasons. First, it is needed to keep the output of the detector steady since the received carrier varies in strength through fading, etc. Secondly, a.g.c. protects the i.f. amplifiers from overloading which causes various forms of distortion. As explained in *Principles and Systems for Radio and TV Mechanics*, a.g.c. is achieved by feeding the d.c. output of the detector back to the r.f. or the first or second i.f. amplifier or both to control their gain. The a.m. detector described above produces a d.c. voltage at its output in the same manner as a d.c. power supply, as shown in Fig.25.2c. The level of this d.c. voltage is determined by the amplitude of the i.f. A high amplitude i.f. produces a high d.c. voltage at the output of the detector and vice versa. The d.c. voltage thus produced is employed as the feedback d.c. control voltage.

Reverse and Forward A.G.C.

There are two types of a.g.c., forward and reverse. Reverse a.g.c. uses the fact that the gain of an amplifier increases as the current through it increases. Conversely, transistors may be manufactured having their gain decreasing as their current increases. These transistors are used in forward a.g.c.

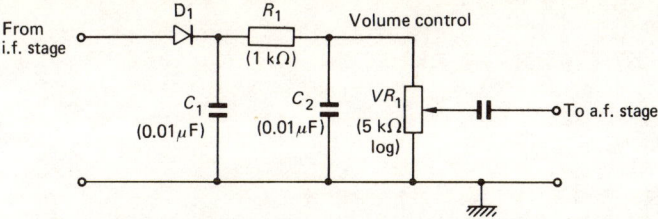

Fig.25.1 Amplitude modulation detector

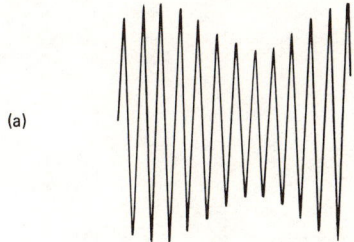

(a)

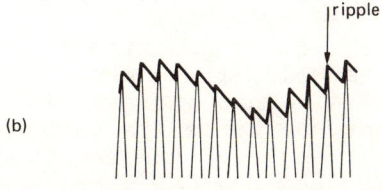

(b)

r.f.
ripple

Fig.25.2
(a) Input to a.m. detector
(b) Effect of time constant $C_1 VR_1$
(c) Output waveform after filter
    circuit
(d) Effect of long time constant

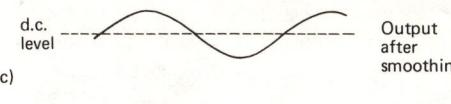

d.c.
level

Output
after
smoothing

(c)

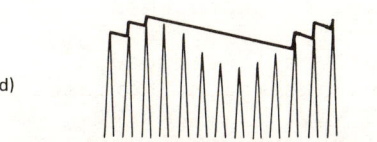

Very long
time constant

(d)

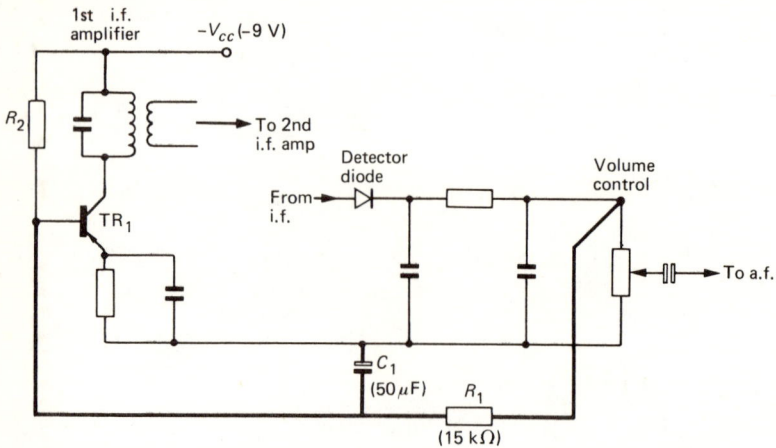

Fig.25.3 Reverse a.g.c. as used in a.m. radio receivers

REVERSE A.G.C. is used in radio receivers and for TV sound i.f. stages. A typical circuit is shown in Fig. 25.3. The a.f. signal is filtered out by low pass filter R_1-C_1 leaving a positive d.c. voltage only, which is used to bias the base of TR_1. If due to bad reception the carrier level and hence the level of i.f. is low, the d.c. output of the detector will be low as well, i.e. less positive making TR_1 base more negative. This increase in the bias of TR_1 increases its current and hence its gain in order to counterbalance the low strength of the received carrier, thus maintaining the output from the detector constant. For strong reception the gain of the first i.f. amplifier decreases, reducing the output from the detector.

FORWARD A.G.C. is used in TV receivers because of its better linearity. Fig.25.4 shows the a.g.c. circuit used in Thorn 1500 TV chassis. The video signal is monitored at the emitter of the video amplifier VT_8. A.g.c. detector diode W_6 is forward biased by negative-going sync pulses to produce a d.c. voltage, which together with bias chain R_1-R_2 biases the a.g.c. amplifier VT_3. C_3 and C_6 are decoupling capacitors for line and field frequencies. The output from VT_3 collector is used to bias VT_5 which in turn biases VT_4 via resistor R_8. A strong signal produces sync pulses with higher amplitudes which when detected by W_6 produces a larger negative d.c. level giving a lower bias to VT_3. VT_3 current is reduced, thereby increasing its collector voltage which increases the forward bias of VT_5 and with it VT_4. VT_5 and VT_4 currents go up and with it their gain goes down, thus maintaining the i.f. level constant.

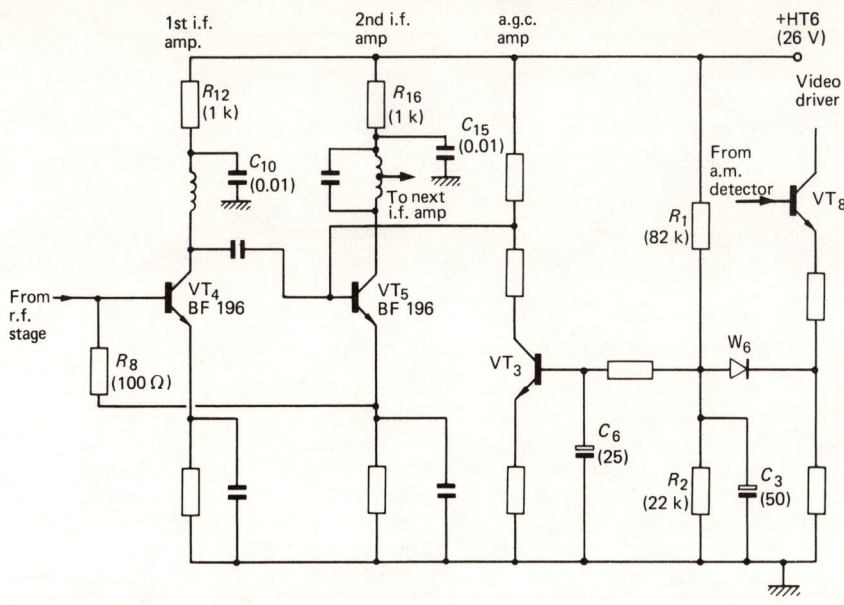

Fig.25.4 Forward a.g.c. as used in TCE 1500 TV receiver (By permission of Thorn Consumer Electronics)

The Frequency Modulation Demodulator or Detector

In frequency modulation (f.m.) the carrier frequency changes or deviates from the centre frequency in accordance with the modulating information. The f.m. detector or discriminator thus have to convert the frequency deviation back to the original information signal.

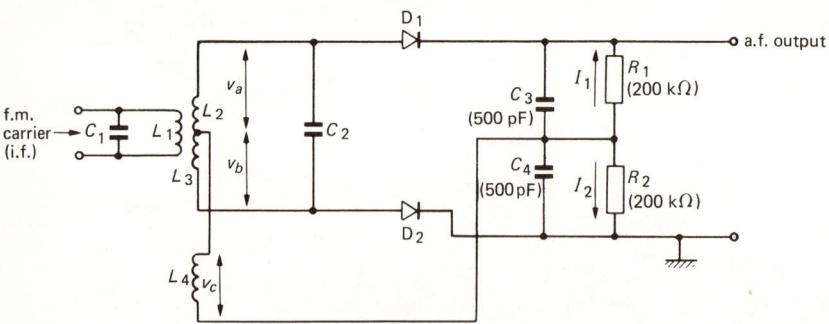

Fig.25.5 Foster Seeley f.m. detector

THE FOSTER-SEELEY DETECTOR
In this detector (Fig.25.5), L_1-C_1 is the tuned circuit
at the output of the last i.f. amplifier. The i.f. is
transformer coupled to L_2-L_3 which is also tuned to the
centre i.f. by capacitor C_2. The centre tap at the secondary produces two voltages in anti-phase to each other. The
voltage v_a across L_2 is fed to the anode of D_1. The
voltage v_c induced in L_4 is fed via C_3 to the cathode of
D_1. Thus v_a and v_c are added up vectorially. The resultant voltage produces a current I_1 through R_1 in the direction shown. Similarly the voltage v_b across L_3 is added
vectorially to v_c. Their resultant produces a current I_2
through R_2 in the opposite direction to I_1. When the
input is unmodulated, i.e. its frequency is at the centre
i.f., v_a and v_b are equal in amplitude and drive equal
currents through R_1 and R_2. Since I_1 flows in the opposite direction to I_2, they cancel each other giving a zero
voltage at the output.
    When the incoming waveform is modulated, its frequency
changes above or below the centre i.f. At frequencies
above the centre i.f., tuned circuit L_2-L_3-C_2 is inductive,
producing a phase shift in v_a and v_b. Voltage v_c on the
other hand suffers no phase shift since L_4 is not tuned.
When v_a and v_b are added separately to v_c, their result-
ants are no longer equal and therefore unequal currents
are produced through R_1 and R_2 resulting in a voltage at
the output. When the incoming frequency is below the
centre i.f., an output voltage is produced in the same way
but with the opposite polarity. Hence the output corresponds to the frequency deviation of the input. Time
constants C_3R_1 and C_4R_2 act as filters to smooth the r.f.
ripple component of the output.

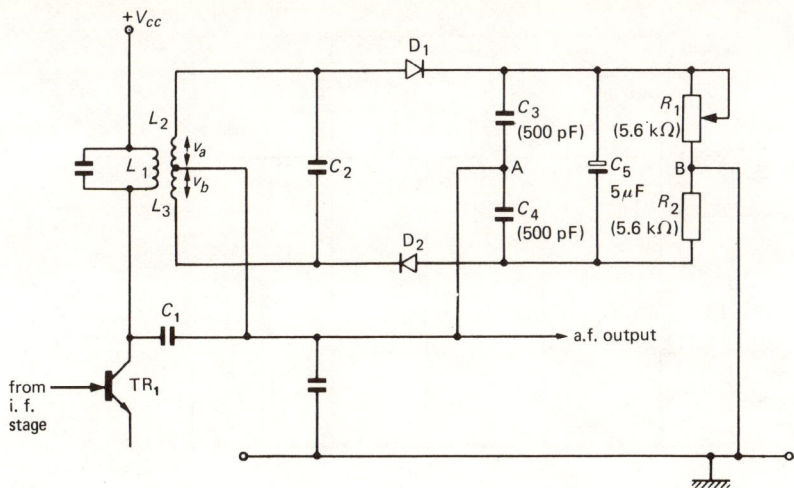

Fig.25.6 Ratio detector

THE RATIO DETECTOR

The Foster-Seeley discriminator suffers from its sensitivity to amplitude changes in the incoming f.m. input. A
limiter has to be used before the discriminator to remove
any amplitude modulation present on the f.m. carrier.
This drawback may be overcome by the use of another type
of detector or discriminator known as the ratio detector,
shown in Fig.25.6. As in the case of the Foster-Seeley,
two signals are added vectorially to produce a resultant.
One signal (v_a or v_b) changes in phase in accordance with
the incoming frequency, while the other (at the collector
of i.f. amplifier TR_1) suffers no such change.

The output is a measure of the p.d. between point A and
point B. At the centre frequency both D_1 and D_2 conduct
equally, producing equal potentials at A and B, and giving
a zero output. When the input frequency deviates from
the centre frequency, the currents through the diodes are
now unequal and an output voltage is produced. The
voltage is, therefore, a measure of the deviation. There
are numerous designs for a ratio detector. That shown in
Fig.25.6 is a common one with C_3 and C_4 as the r.f. and
C_5 as the a.f. decoupling capacitors. R_1 is made variable
to ensure balance in the circuit.

Note that compared with the Foster-Seeley discriminator,
the ratio detector has two diodes connected "opposite to
each other".

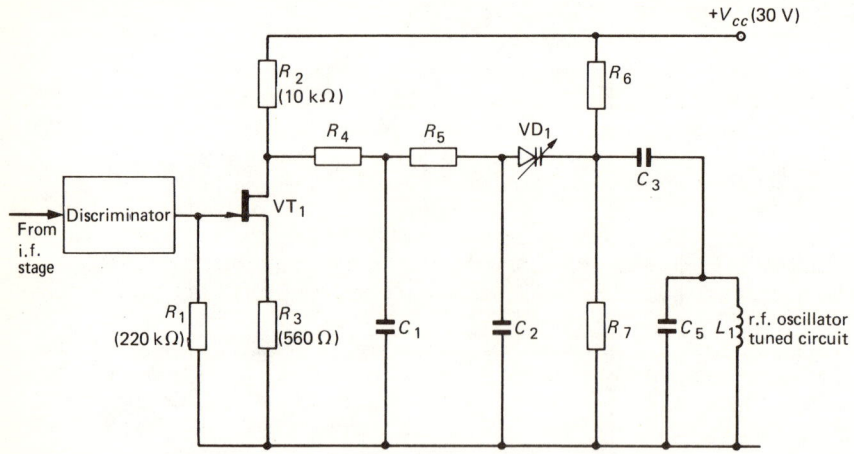

Fig.25.7 A.f.c. discriminator with fet d.c. amplifier VT$_1$

Automatic Frequency Control

As explained in *Principles and Systems for Radio and TV Mechanics*, automatic frequency control (a.f.c.) is used in f.m. radio, as well as TV receivers, to ensure a stable intermediate frequency. The i.f. is monitored by a discriminator which produces a d.c. control voltage which is used to determine the frequency of the local oscillator.

Fig.25.7 shows an a.f.c. discriminator together with a d.c. amplifier. The discriminator produces a d.c. output which corresponds to the error in the i.f. frequency. This d.c. control voltage is amplified by VT$_1$ and smoothed by R_4-C_1 and R_5-C_2. Varicap diode VD$_1$ is reverse biased by R_6-R_7. Its reverse bias is varied by the d.c. control voltage going into its anode, thereby changing its capacitance which then changes the frequency of oscillation of L_1-C_5 in such a way as to correct the original error.

26 Logic Gates and Counters

Logic circuits are used to perform many functions in electronic equipments such as computers, calculators, process control systems. These circuits are constructed using basic units known as logic gates. In *Principles and Systems for Radio and TV Mechanics,* logic gates, their types and functions were introduced. In this book we are concerned with their construction. Diodes, transistors (bipolar and unipolar) as well as valves may be used to build up gates circuits.

When employed in gate circuits, diodes and transistors are used as fast switching devices. A diode when reverse (or zero) biased acts as an open circuit similar to an open switch. When forward biased it acts as a short circuit similar to a closed switch. The forward voltage drop (0.6 V for silicon) is small and may be neglected.

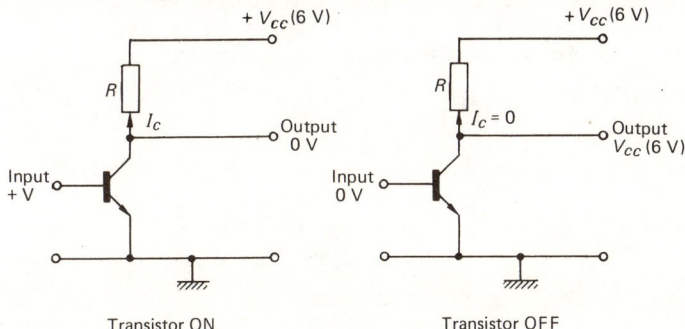

Fig.26.1 The transistor as a switch

Switching transistors are used in logic circuits for their fast switching action. When the b-e junction of a bipolar transistor is forward biased the transistor is on (saturated), acting as a short circuit (or a closed switch) with collector potential down to emitter (Fig.26.1). With reverse or zero bias, the transistor is off, acting as an open circuit with the collector potential at V_{cc}.

Positive and Negative Logic

In logic there are only two states or levels: 0 and 1. Logic-0 corresponds to zero volts and logic-1 to a specified positive or negative voltage. For positive logic, logic-1 corresponds to a positive voltage (e.g. +6 V). For negative logic, logic-1 corresponds to a negative voltage (e.g. -6 V). In the rest of this section, positive logic will be assumed unless otherwise stated.

Logic gates may be classified according to the components used. For example, diode-resistor logic uses diodes and resistors, and so on.

Diode-Resistor Logic (DRL)

OR gate A basic two input OR gate is shown in Fig. 26.2 and consists of two diodes and a resistor. Positive logic is used with logic-0 = 0 V and logic-1 = 6 V. With both inputs at logic-0 level (i.e. both anodes connected to chassis), the diodes are zero biased and current through the resistor is zero giving an output of 0. When input A is at logic-1 level (i.e. D_1 anode connected to a 6 V rail), D_1 is forward biased passing current through R giving an output of 1 (approximately 6 V). When input B is at logic-1, D_2 conducts and output is also 1. Similarly, when both inputs are at logic-1 both diodes conduct and output remains at 1. The truth table for the gate is shown in Fig.26.2. Typical input and output waveforms are shown in Fig.26.3. Note that an output of 1 is present when there is an input at either A or B.

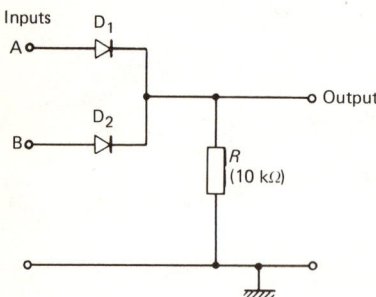

A	B	o/p
0	0	0
1	0	1
0	1	1
1	1	1

Fig.26.2 OR gate
        (a) DRL circuit, (b) truth table

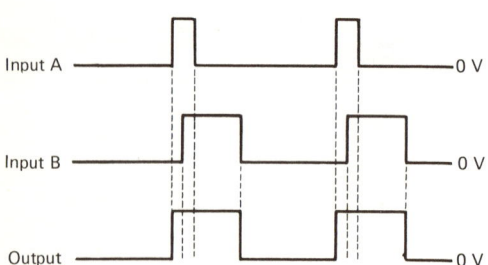

Fig.26.3 Input and output pulses of a 2-input OR gate

AND gate   In the gate shown in Fig.26.4, the anodes of
D_1 and D_2 are connected towards a positive potential
(10 V) via load resistor R_1. When both inputs are at
logic-0, D_1 and D_2 are forward biased and act as a short
circuit.   Hence the anodes will be at the same potential
as the cathodes, namely zero.   When A is 1 (say 6 V) and
B is 0, D_2 remains forward biased and output remains at
0, giving an output of 0, D_1 however is reverse biased.
When both inputs are at 1, say 6 V, then both are forward
biased and the output at their anodes is at the same
level as the input at the cathode, namely 6 V or logic-1.
Typical input and output waveforms in Fig.26.5 show an
output of 1 when both inputs are 1.

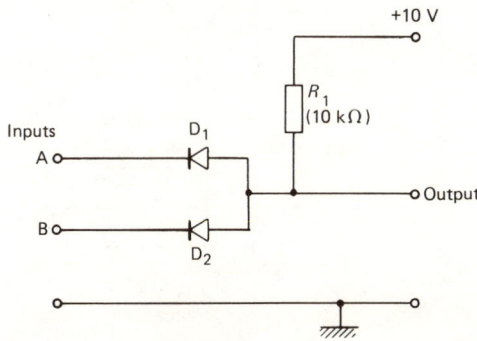

A	B	o/p
0	0	0
1	0	0
0	1	0
1	1	1

Fig.26.4 AND gate
         (a) DRL circuit, (b) truth table

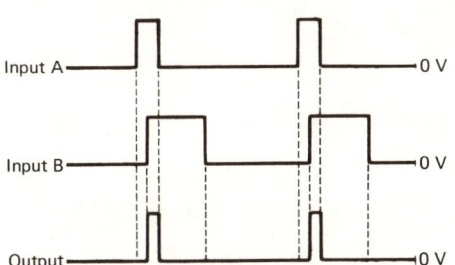

Fig.26.5 Input and output pulses of a 2-input AND gate

Resistor-Transistor Logic (RTL)

NOT gate A simple transistor inverter may act as a NOT gate as shown in Fig.26.6. R_1 is a limiter resistor for the base current and R_2 is the load. With a 0 input, TR_1 is off and the collector is at V_{cc} or logic-1. When the input is at 1, say 6 V, TR_1 is ON (saturated) and collector bottomed with output at 0. To ensure that the transistor remains OFF with 0 input, the base is taken to a negative potential $-V_{BB}$ as shown. Typical waveforms are shown in Fig.26.7.

Note that a NOT gate has one input only. When several inputs are used it becomes a NOR gate.

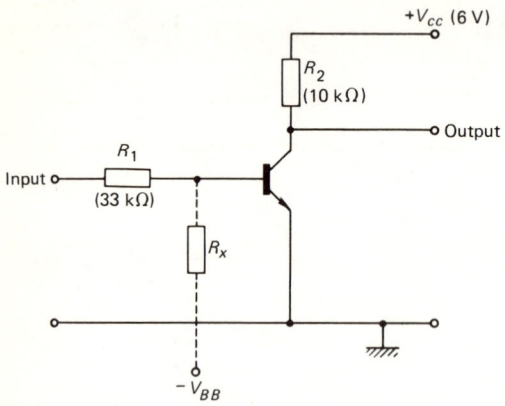

i/p	o/p
1	0
0	1

Fig.26.6 NOT gate
(a) RTL circuit, (b) truth table

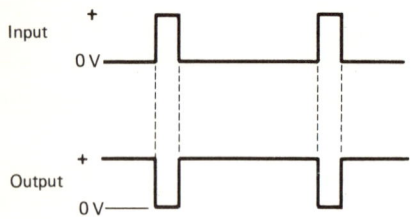

Fig.26.7 Input and output pulses of a NOT gate

NOR gate   When both inputs in Fig.26.8 are at 0, TR_1 is cut-off, giving a 1 output. When either or both inputs are 1, TR_1 saturates giving a 0 output. C_1 and C_2 are speed-up capacitors. Typical waveforms are shown in Fig. 26.9.

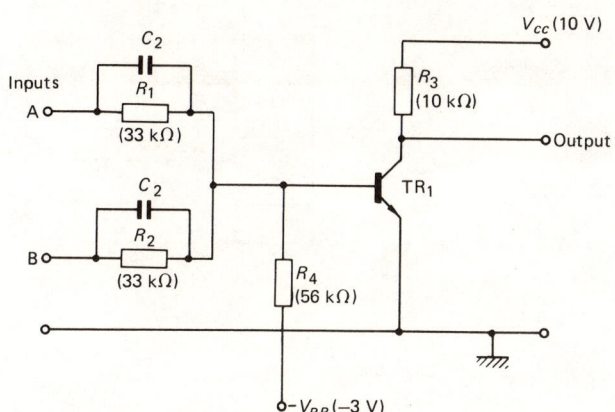

A	B	o/p
0	0	1
1	0	0
0	1	0
1	1	0

Fig.26.8 NOR gate
(a) RTL circuit, (b) truth table

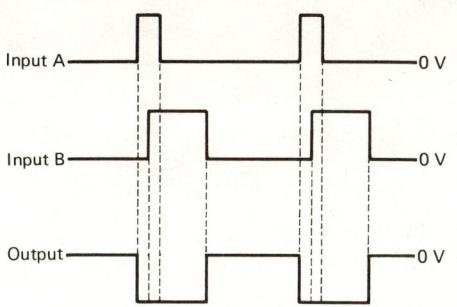

Fig.26.9 Input and output pulses of a 2-input NOR gate

NAND gate   The circuit in Fig.26.10 employs two transistors in series. When either transistor is off, no current flows through R_3 and output is 1. When and only when both inputs are 1, TR_1 and TR_2 saturate and output is 0. Typical waveforms are shown in Fig.26.11.

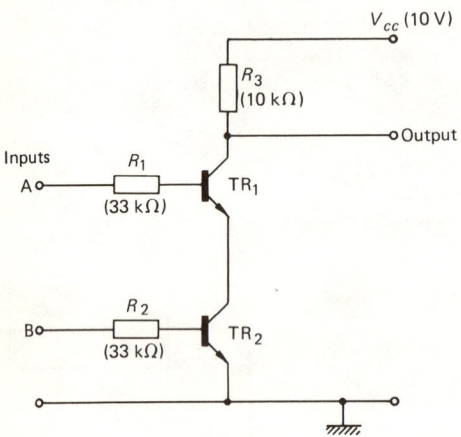

A	B	C
0	0	1
1	0	1
0	1	1
1	1	0

Fig.26.10 NAND gate
     (a) RTL circuit, (b) truth table

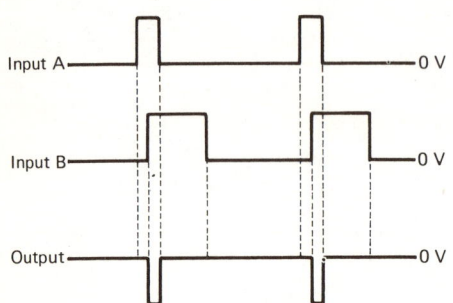

Fig.26.11 Input and output pulses of a 2-input NAND gate

NOR gate   Fig.26.12 shows a NOR gate using two pnp transistors employing negative logic. When both inputs are at 0, both transistors are off and output is 1. When one or both inputs are at logic-1 level, say -6 V, one or both transistors saturates, giving an output of 0. A common emitter resistor R_4 is used to improve the d.c. stability of the gate.

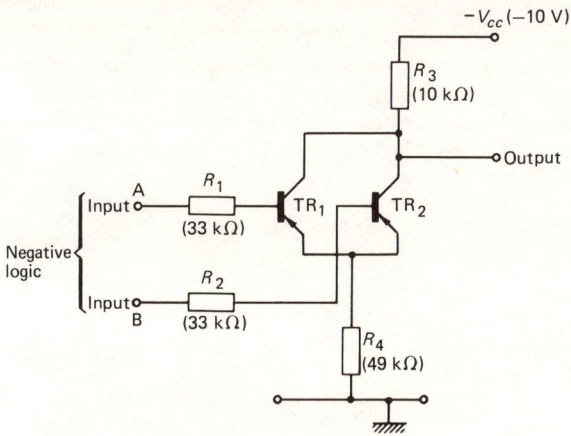

Fig.26.12 NOR gate (RTL circuit)

Diode-Transistor Logic (DTL)

A basic form of a DTL NAND gate is shown in Fig.26.13. It
consists of an AND gate followed by a NOT gate making a
NAND gate. One problem that is encountered in DTL gates,
such as the one under consideration, is that with one or
both inputs at 0, the output from the AND section at point
P has a voltage of 0.6 V, the forward voltage drop of D_1
and D_2. Although this voltage still gives logic-0 it is
nonetheless enough to forward bias TR_1. To avoid this, D_3
and D_4 called voltage offset diodes are connected as shown.
This way TR_1 will conduct only when potential at P is
high enough to forward bias D_3 and D_4 as well as the b-e
junction of TR_1, a voltage of approximately 1.8 V.

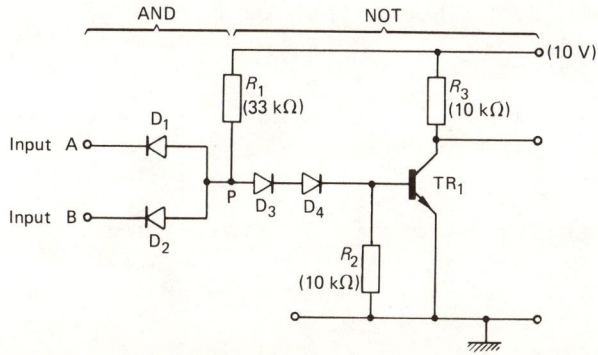

Fig.26.13 NAND gate (DTL circuit)

Transistor-Transistor Logic (TTL)

This type of logic is used in integrated circuits and has the advantage of fast switching action. Fig.26.14 shows a NAND gate using a multiple emitter transistor TR_1 at the input. With both inputs at logic 0, TR_1 saturates, making its collector go down to approximately 0 V. Hence TR_2 is off and the output is 1. With either or both inputs at logic 1, TR_1 turns off switching TR_2 on to give an output of 0.

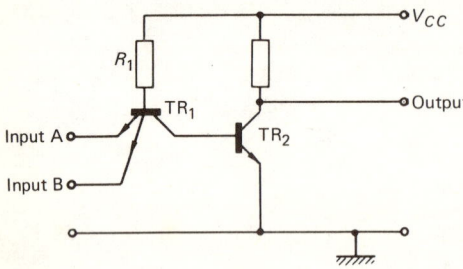

Fig.26.14 NAND gate (TTL circuit)

FET Logic Gates

Most logic gates are now manufactured as integrated circuits, ICs. A large number of gates may then be constructed on a tiny chip of silicon of 1 mm × 2 mm. Because of their simplicity fets are more commonly used than bipolar transistors. Fig.26.15 shows a NOR gate similar in principle to that shown in Fig.26.12 operated by negative logic. VT_1 and VT_2 are p-channel mosfets of the normally-off (enhancement) type. When both inputs are at logic-0 level, VT_1 and VT_2 are cut off giving an output of 1 ($-V_{DD}$ = -20 V). When either or both inputs are at logic-1 level (say -20 V) one or both transistors are on, giving an output of 0.

Fig.26.16 shows a NAND gate similar in principle to that in Fig.26.10 using an n-channel normally-off mosfet. Positive d.c. supply is used and positive logic is necessary to operate the gate. VT_3 is biased permanently on by taking its gate to V_{DD} and acts as a load to the gate. When one or both inputs are at 0 level, one or both transistors are off giving a 1 output. Only when both inputs are 1 will current flow giving an output of 0.

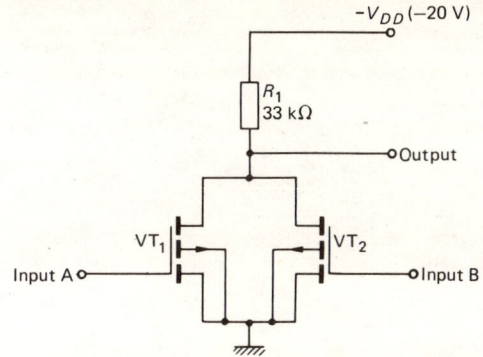

Fig.26.15 NOR gate (fet logic)

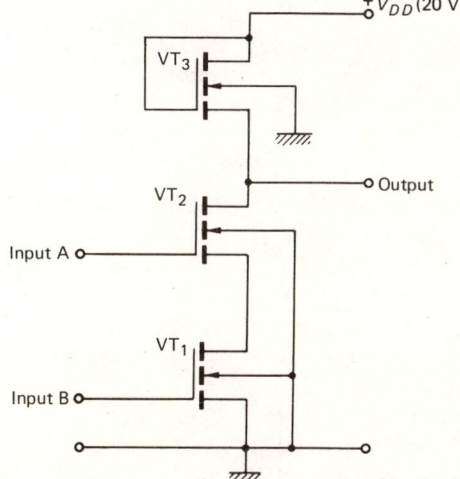

Fig.26.16 NAND gate (fet logic)

Counters

The basic binary counter is a bistable or ÷2 device also
known as a FLIP-FLOP. A number of these elements con-
stitute a counter. Fig.26.17 shows a binary counter
consisting of three flip-flops. Each flip-flop is capable of dividing incoming pulses by a factor of 2. Two
successive flip-flops with thus divide by 2 × 2 = 4 (÷4),
and three flip-flops divide by 2 × 2 × 2 = 8 (÷8). In
other words, for every 8 input pulses, 4 appear at output
A, 2 at output B, and 1 at output C, as shown in Fig.
26.17.

As explained in section 22, only one edge of the input
pulse causes the bistable to change its state. The input
changes from O to 1 and back to O and so on for every
pulse. It is normally assumed that the negative-going
edge of the pulse (i.e. from 1 to O) changes the state of
the flip-flop.

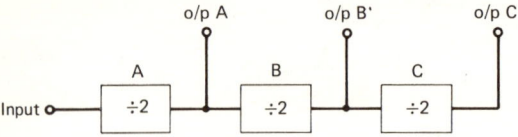

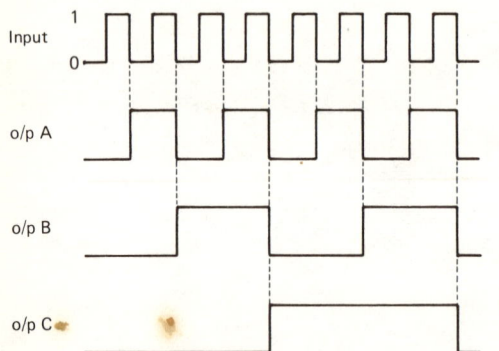

Fig.26.17 *3-element binary counter*

Table 26.1 gives the output from each flip-flop element for eight input pulses.

TABLE 26.1

Pulse	o/p A units (2^0)	o/p B twos (2^1)	o/p C fours (2^2)
0	0	0	0
1	1	0	0
2	0	1	0
3	1	1	0
4	0	0	1
5	1	0	1
6	0	1	1
7	1	1	1
8	0	0	0

(repeat)

Output A from bistable A represents the 2^0 or units column, output B the 2^1 or twos column, and output C the 2^2 or fours column. Hence after 6 pulses the outputs are as follows: A (units) is 0, B (twos) is 1, and C (fours) is 1, making 0 + 2 + 4 = 6. In binary coding this is written as 110 in the order CBA. It should be noted that the units column is the first bistable starting from the input. In binary coding on the other hand the units column is always on the right-hand side.

At the seventh pulse all outputs are 1. Pulse number 8 resets all flip-flops back to 0.

Note that the output of each flip-flop represents a column in the binary code. The binary code itself is in the order of CBA.

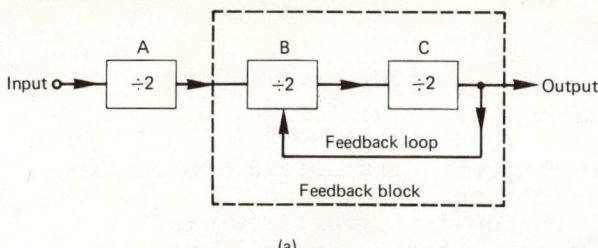

(a)

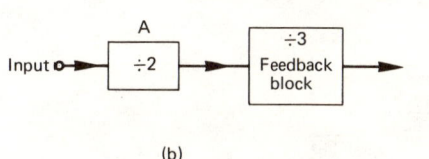

(b)

Fig.26.18
(a) ÷6 counter employing feedback
across two bistables B and C
(b) reduces the feedback loop to
a ÷(4-1), i.e. ÷3 block

Feedback

Feedback may be applied to binary counters in order to change the factor by which the input is divided. For example consider the feedback applied to the 3-element counter shown in Fig.26.18a. The count proceeds normally up to the third pulse when outputs CBA are 011 (see Table 26.2). The fourth pulse changes A to 0, B to 0, and C to 1. Without feedback output CBA would be 100. However with feedback the change in C is fed back to B changing it back to 1 giving CBA reading of 110. Pulse 5 then produces a reading of 111 and the 6th resets all elements back to 0. Thus we have a ÷6 counter.

TABLE 26.2

Pulse	o/p A	o/p B	o/p C	
0	0	0	0	
1	1	0	0	
2	0	1	0	
3	1	1	0	repeat
Feedback →	(0)	(0)	(1)	
4	0	1	1	
5	1	1	1	
6	0	0	0	

In general it can be shown that a feedback loop reduces the division factor of the bistables inside the loop by 1. Taking the above example, the bistables inside the feedback loop are B and C. Without feedback they divide by a factor of 4. With feedback, bistables B and C form a feedback block having a division factor of 4 - 1 = 3 as shown in Fig.26.18b. With bistable A outside the feedback block the total factor of division is 2 × 3 = 6.

The Decimal Counter

Fig.26.19a shows a decimal counter employing two feedback
loops. Two feedback blocks are thus formed: block 1
enclosing bistables C and D, dividing by 4 - 1 = 3; block
2 encloses block 1 together with bistable B. Referring
to Fig.26.19b, without feedback block 2 would divide by
2 × 3 = 6. With feedback the division factor is 6 - 1
= 5. With bistable A outside the feedback loops, the
total division factor is thus 2 × 5 = 10.

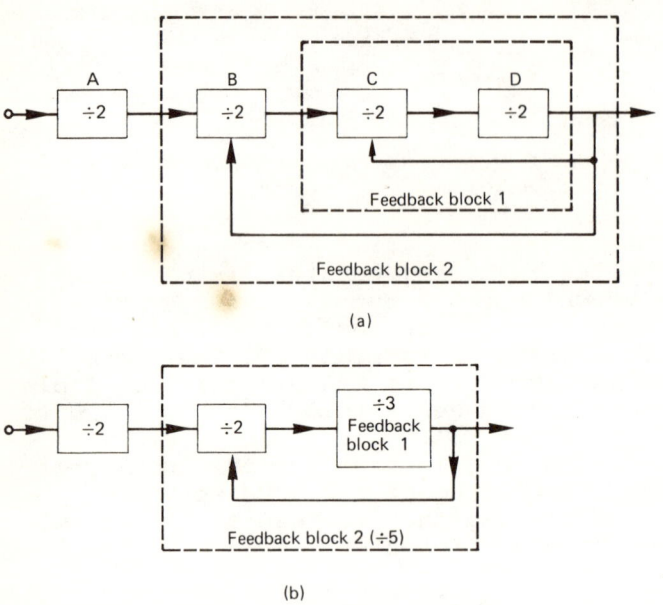

(a)

(b)

Fig.26.19 (a) Decimal counter
(b) Reduction to feedback blocks

Shift Register

To transfer data from one part of a system such as a
computer to another, two methods are possible. The first
and quickest is to transfer all the digits in one operation. Thus, for transferring eight digits, eight separate
lines are required. For distances up to few metres this
method is used, but for longer distances such as data
transmission between cities this method becomes too expensive. A second but slower method is used. The data
information are transmitted digit by digit along a single
wire. A shift register may be used to move the binary
digits one stage to the left or to the right. The shift
register consists of a number of flip-flops capable of
rearranging the binary digits in successive order.

Ring Counter

A ring counter consists of a standard counter comprising a number of bistables in which the output is taken back to the input - hence its name. Pulses may thus circulate round the counter from the input to the output and back to the input and so on. At the end of each cycle the output may be taken to trigger another counter. For example the output of a scale-of-ten counter may be used to trigger another scale-of-ten counter to make a scale of 100.

27 Relays

BPO and Reed Relays

Relays are electromagnetic devices used to break, make or change over one or more contacts. They are extensively used in telephone switching systems. Compared with electronic switching, e.g. a transistor, relays are very slow but they have the advantage of being able to operate more than one contact at once.

The most commonly used relay is the British Post Office relay shown in Fig.27.1. It consists of a coil which when magnetised by a d.c. current attracts a metal strip known as the armature. The movement of the armature is then used to operate the contacts shown. The current through the coil must create a large enough magnetic field to overcome the spring tension holding the two contacts in their normal position.

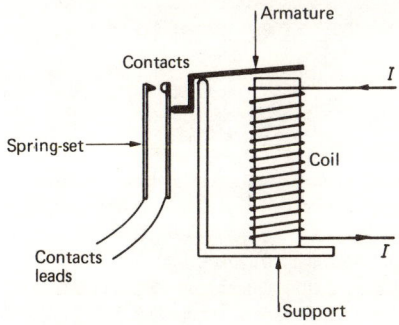

Fig.27.1 Post Office Relay

The speed of operation of the post office relay is slow
due to the inertia of the moving parts as well as to the
time lag caused by the delay in the growth of current
through the inductive coil. A fast switching relay is
the REED RELAY shown in Fig.27.2. The contacts are very
thin and of low mass. They are sealed in a glass tube
which may be filled with an inert gas to avoid oxidisation
of the contacts. The contacts shown open may be closed
by a magnetic field created by a current passing through
the coil.

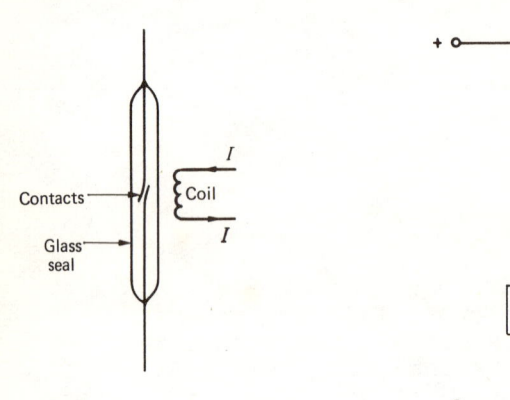

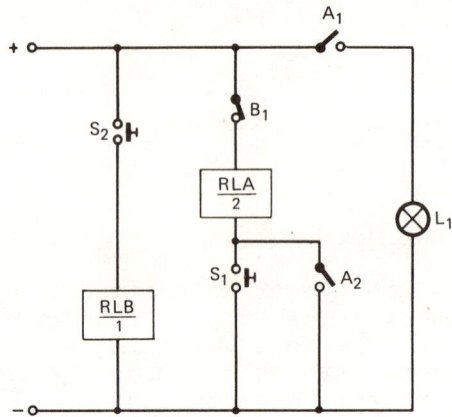

Fig.27.2 Reed Relay   *Fig.27.3 Relay circuit used to switch*
                              lamp L_1 on and off

NOTATION Since one relay coil may be used to operate a
number of contacts, each contact is given a notation
referring it to the coil responsible for its operation.
Fig.27.3 shows a relay circuit using the standard notation. S_1 and S_2 are two push button switches, i.e. they
make when pressed but break as soon as they are released.

$\dfrac{RLA}{2}$ is relay coil A operating two contacts A_1 and A_2
which are normally open.

$\dfrac{RLB}{1}$ is relay coil B operating one contact B_1 which is
normally closed.

When S_1 is closed, relay A is operated closing A_1 and
A_2. Lamp L_1 therefore goes on. The lamp stays on even
when S_1 is released since contact A_2 is now closed which
keeps RLA operated. To switch the lamp off, S_2 is pressed.
This operates relay B which opens contact B_1 releasing
relay A which opens contacts A_1 and A_2. The lamp remains
off when S_2 is released.

Transistor-operated Relay

A relay may be operated by a transistor as shown in Fig. 27.4. When switch S is closed, TR_1 conducts passing current through the relay coil which operates the contacts.

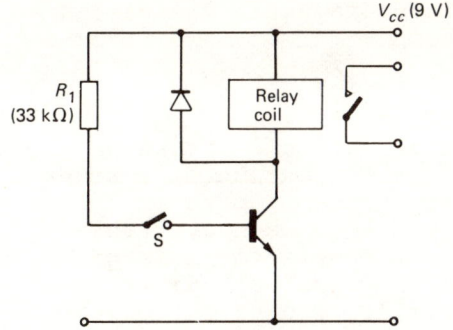

Fig.27.4 Transistor operated relay

A diode D_1 is connected across the relay coil to protect the transistor from the back e.m.f. in the coil when the current through it decays (as was explained for the case of the blocking oscillator, p.144).

Slugging

On some occasions a delay in the operation of a relay is required. This is known as slugging the relay and may be achieved by placing a copper ring round the core of the relay coil. Slugging may also be achieved as shown in

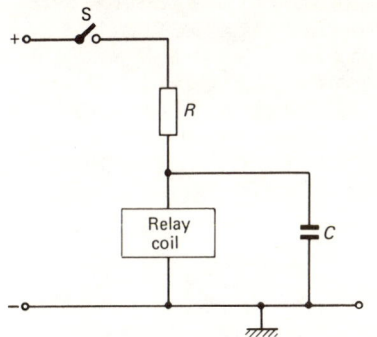

Fig.27.5 Slugging a relay. Both the operate and the release functions are delayed

Fig.27.5. When switch S is closed, capacitor C charges up slowly through resistor R. The voltage and hence the current growth in the coil is slowed down, thus delaying the "operate" function. When S is opened, C discharges slowly through R, thus delaying the "release" function of the relay.

28 Transmission Lines, Propagation and Aerials

Having produced electric energy, some means must be found
for sending or transmitting it from its source to the
user.   The answer is that transmission lines may be used
to send energy by cable, and electromagnetic radiation
may be used for radio transmission.

Transmission Lines

Cables used to carry high frequency signals are known as
transmission lines.   The signal at the output of a transmission line varies from that at its input both in
amplitude and phase.   To represent this change, transmission lines may be considered to consist of a large
number of resistors, inductors and capacitors distributed
along the line as shown in Fig.28.1.   The effect of the

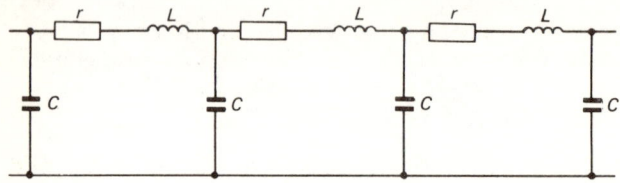

Fig.28.1 A transmission line represented by r, L and C

resistor is to attenuate the signal amplitude.   The amount
of attenuation is the same regardless of the frequency.
It is normally small and has to be taken into account for
very long lines only.   At low frequencies the effect of
inductor L and capacitor C is negligible.   At high
frequencies however their effect becomes noticeable even
for a short length of line.   The transmission line represents an impedance to incoming signals, the value of which
depends on the frequency as well as the length of the line.
At radio frequencies a transmission line may be used as a
resonant circuit.

MATCHING THE LINE
Consider a transmission line of infinite length.   Looking
at the transmitting end, the line presents an impedance
known as the characteristic impedance Z_0.   The value of Z_0
depends on the size of the conductor, the spacing, and
insulation between live and earth.

For a short line terminated by an impedance other than its characteristic impedance, energy transmitted from one end travels down the line, and gets reflected at the terminating impedance and a standing wave is formed along the line. The energy absorbed by the load is therefore low since some energy is reflected back to the transmitting end. However, if we now terminate the line with its characteristic impedance Z_0, the transmitted energy wave sees an infinitely long line and reflection ceases. All the energy is thus absorbed by the load, and energy or power transfer from transmitter to load is at a maximum. This is known as matching the transmission line. Fig.28.2 shows matching arrangement necessary for maximum power transfer between transmitter and receiver using a transmission line.

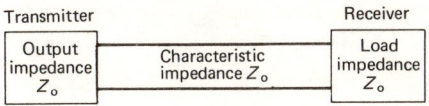

Fig.28.2 Matching between a transmitter and a receiver using a transmission line

THE COAXIAL CABLE
This type of cable is the most popular type for u.h.f. and v.h.f. applications. It has a characteristic impedance of 75Ω. It consists of an inner conductor surrounded by an

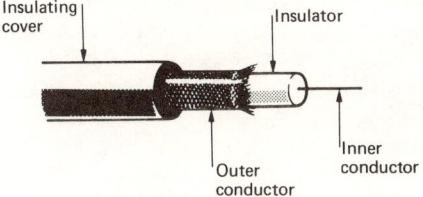

Fig.28.3 Construction of a coaxial cable

outer conductor in the form of a tube normally of woven wire braid as shown in Fig.28.3. The outer conductor is normally used as earth which then functions as a screen reducing any interference due to stray radiation.

Radio Propagation

Radio transmission uses electromagnetic waves which travel through space at the speed of light (300 000 000 m/s). Since electromagnetic waves travel in straight lines, a receiving aerial must be within sight of a transmitting aerial for transmission to be received. Hence transmitters are placed on high ground to cover a large area.

Very long distances may be covered by refraction of
radio waves in the ionsphere. The ionsphere is an ionised
gas layer in the upper atmosphere, about 200 km above the
earth's surface. Its effect is to refract or bend electromagnetic waves passing through it. If refraction is high
enough, radio waves travelling into the ionsphere from
earth may travel back to earth as shown in Fig.28.4. The

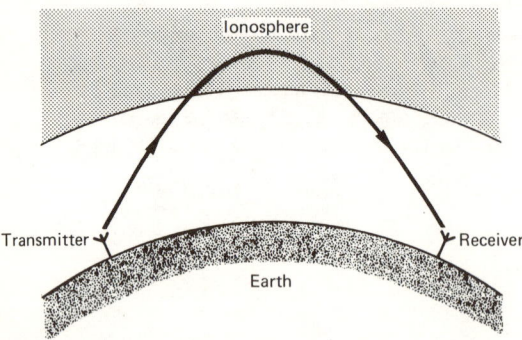

Fig.28.4 Refraction of radio waves in the ionosphere

height and thickness of the ionsphere varies from day to
day, day and night, place to place, and with atmospheric
conditions. Hence the variation in reception that occurs
from day to day, etc., for long distance transmissions.
  The refraction of radio waves at the ionsphere gets
smaller as the frequency increases. It is very effective
and hence widely used for medium and short waves. At
v.h.f., u.h.f. or higher frequencies, the refraction is
insufficient and other methods such as satellites are
used to cover long distances.

Aerials

Aerials are used both for transmitting a radio wave and
for receiving it. The principle of operation of a transmitting aerial is that, when current flows in a short
straight conductor or aerial, energy in the form of an
electric field and a magnetic field is radiated.
  A vertical aerial creates a vertical electric field
(with a horizontal magnetic field) and the radiation is
said to be vertically polarized. The electric field is
at right angles to the magnetic field with the direction
of propagation at right angles to both as shown in Fig.
28.5.
  A horizontal aerial creates a horizontal electric field
(with a vertical magnetic field) and the radiation is
said to be horizontally polarized.

The principle of operation of a receiving aerial is
that if a conductor (or aerial) is placed in the path of
electromagnetic radiation, current is induced in the
conductor.

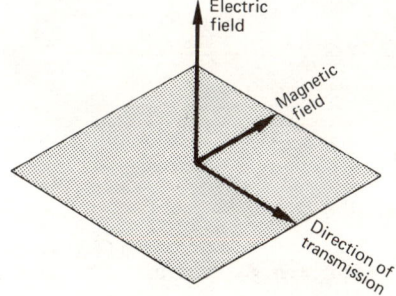

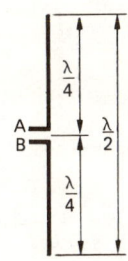

Fig.28.5 The electromagnetic wave   *Fig.28.6 The half wavelength*
dipole

For maximum radiated energy from a transmitting aerial
and maximum received energy by a receiving aerial, the
aerial has to be of the correct length. The optimum
length is determined by the wavelength of the signal.
The wavelength of a radio wave is the distance covered by
one cycle of the waveform and thus depends on its frequency
and speed of travel.

$$\text{Wavelength} = \frac{\text{Velocity of radio waves } c}{\text{Frequency } f}$$

where c is the speed of light.
The basic aerial used for v.h.f. and u.h.f. applications
is the half wavelength ($\lambda/2$) dipole. It consists of two
conductors or rods, a quarter wavelength ($\lambda/4$) each, as
shown in Fig.28.6. For a transmitting aerial, r.f. power
is fed into A and B, producing electromagnetic radiation.
A receiving $\lambda/2$ dipole placed in the path of the radiation would have r.f. voltage induced across A and B which
is fed to the receiver. For maximum reception the
receiving dipole must be correctly placed with respect to
the incoming wave. For vertically polarized radio waves,
for example, the receiving aerial must be vertically
placed.

DIRECTIONAL AERIALS To improve transmission and reception, directional aerials employing reflectors and directors are used. For a transmitting aerial they have the
effect of concentrating or emphasising the radiation in
one direction. For a receiving aerial they have the
effect of increasing the strength of the received signal.
Since directional aerials for transmission and reception
are identical, reference will be made to receiving aerials
only.

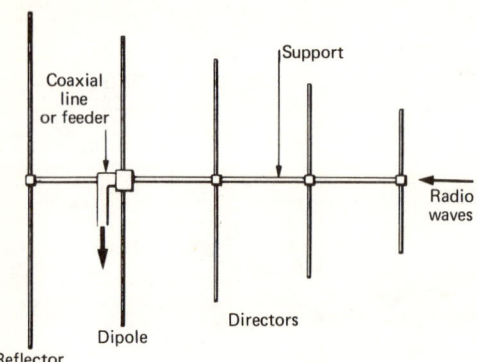

Fig.28.7 Directional aerial commonly used for v.h.f. and u.h.f. receivers

A reflector may be just a conducting rod slightly longer than λ/2 length placed approximately λ/4 behind the dipole as shown in Fig.28.7. Its function is to reflect radio waves back to the dipole aerial.

Directors are conducting rods slightly shorter than λ/2 in length placed in front of the aerial as shown in Fig. 28.7. Their function is to direct the radiation in the vicinity towards the dipole. Note that the reflector and directors are not connected electrically to the dipole.

29 Time Delay Circuits

It is necessary sometimes to delay the application of a pulse or to increase its width. One simple method of delaying a pulse is to make use of the charging time of a capacitor through a resistor. Such a circuit is used to slug a relay as explained on p.171. The same method can be used to delay the switching of a transistor.

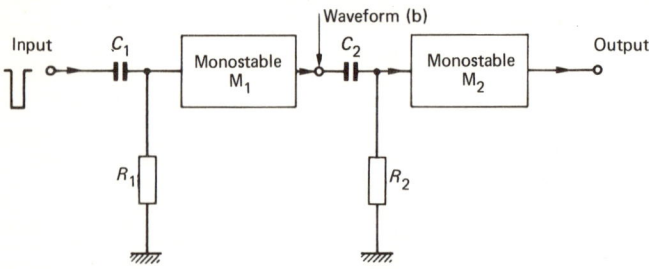

Fig.29.1 Time delay circuit using monostables

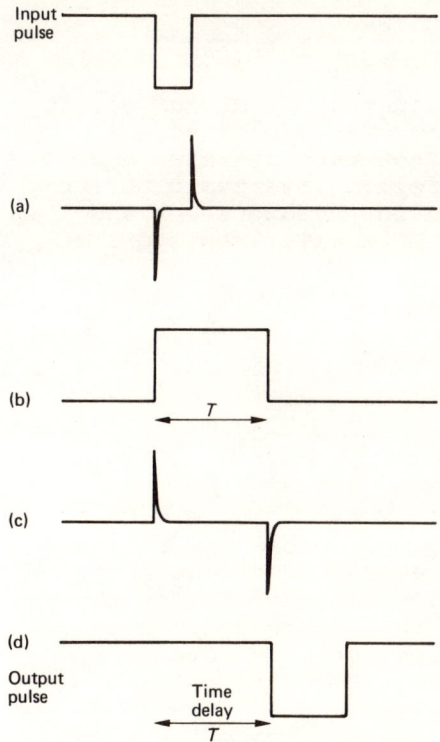

Input
pulse

(a)

(b) T

(c)

(d)

Output
pulse   Time
delay
T

Fig.29.2 Waveform of the circuit in Fig.29.1

Monostable multivibrators may be used to widen a pulse or to delay its action. Fig.29.1 shows a block diagram for a delay circuit using two monostables M_1 and M_2. A square wave pulse input is differentiated by C_1-R_1 to produce negative and positive going pulses as shown in Fig.29.2a. Assuming that a negative going pulse is required to switch the monostable over, then the output will be as shown in 29.2b. The width of the pulse is determined by the time constant T of the monostable and may be made longer than the width of the input pulse.

To delay the application of the pulse another monostable M_2 is used. The pulse shown in 29.2b is differentiated by C_2-R_2 to produce the positive and negative spikes shown in 29.2c. The negative spike is once again used to switch the monostable over producing the square wave shown in 29.2d. Compared with the input pulse, the output is delayed by time T determined by the time constant of the first monostable M_1. Note that the time constant of monostable M_2 determines the width of the output pulse only, and not the delay time.

30 Measuring Devices

In *Principles and Systems for Radio and TV Mechanics*, the operation of the moving coil meter was explained. A deflection is produced by a current passing through a coil placed inside a magnetic field. This basic deflection may have its scale extended for both current and voltage by the use of shunt and multiplier resistors respectively.

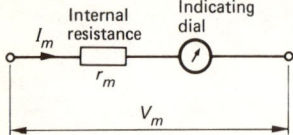

Fig.30.1 Basic moving coil meter

The basic meter (Fig.30.1) may be represented by resistance r_m known as the internal resistance of the meter and an indicating dial. For full-scale deflection the meter current I_m is known as the full-scale deflection current I_{fsd} which passing through meter resistance r_m produces full-scale deflection voltage

$$V_{fsd} = I_{fsd} \times r_m$$

The full-scale deflection current gives a measure of the sensitivity of the meter. A meter that requires a small current to give full-scale deflection is more sensitive than a meter that requires a larger current for full-scale deflection. Hence

$$\text{Sensitivity} = \frac{1}{\text{Full-scale deflection current}} = \frac{1}{I_{fsd}} \ \Omega/V$$

Shunt Resistor

To extend the *current* range of a meter, a shunt resistor R_s is added across the meter as shown in Fig.30.2. The effect of the shunt resistor is to direct part of the total current I_T away from the basic meter and into the shunt resistor.

Given that the full-scale deflection current I_{fsd} = 20 mA and that the meter has a resistance r_m = 5Ω, then the value of the shunt resistor required to extend the range to 200 mA can be calculated as follows.

At full-scale deflection, total current I_T = 200 mA, meter current $I_m = I_{fsd}$ = 20 mA. Hence shunt current I_s = 200 - 20 = 180 mA. But

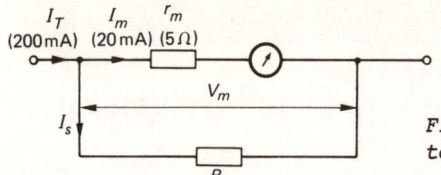

Fig.30.2 *The use of a shunt resistor R_s to extend the current range of a meter*

Voltage across shunt resistor = Voltage across the
basic meter V_m

$$V_m = V_{fsd} = I_{fsd} \times r_m = 20 \text{ mA} \times 5 = 100 \text{ mV}$$

Therefore

$$R_s = \frac{V_{fsd}}{I_s} = \frac{100 \text{ mV}}{180 \text{ mA}} = 0.556\Omega$$

The total resistance of the meter is now smaller than that
of the basic meter.

$$\text{Total resistance} = \frac{5 \times 0.556}{5 + 0.556} = 0.5\Omega$$

To extend the range further, say to 300 mA, a shunt
resistor of 100 mV/280 mA = 0.357Ω is required and so on.

The Multiplier

To extend the *voltage* range of a meter, a series resistor
known as a multiplier is used as shown in Fig.30.3. A
large portion of the total voltage develops across the
multiplier leaving a smaller portion across the meter
which is used to produce a deflection. Note that the only
current through the multiplier is the meter current I_m

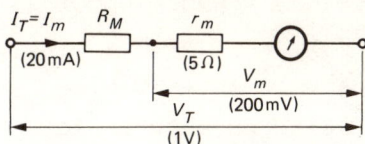

Fig.30.3 *The use of a multiplier R_M to extend the voltage range of a meter*

itself. For the same basic meter as in the previous
example where $I_{fsd} = 20$ mA, $r_m = 5\Omega$ and a full-scale
deflection voltage $V_{fsd} = 20 \times 5\Omega = 200$ mV, then to
extend the voltage range of the meter to 1 V (= 1000 mV)
the value of the multiplier can be calculated as follows.

At full-scale deflection, total voltage = 1000 mV

Voltage across the multiplier $V_R = V_T - V_{fsd}$
$$= 1000 - 200 = 800 \text{ mV.}$$

Current through the multiplier = $I_{fsd} = 20$ mA

Hence the multiplier resistance $R_M = \frac{V_R}{I} = \frac{800 \text{ mV}}{20 \text{ mA}} = 40\Omega.$

180

The total resistance of the meter is now greater than the original basic meter.

Total resistance = 40 + 5 = 45Ω.
To extend the range of the meter to say 10 V, a multiplier of 9800 mV/20 mA = 490Ω is required, and so on.

Examples

Example (1)   Fig.30.4 shows a range selector network for a voltmeter.   The basic meter has a resistance r_m = 50Ω

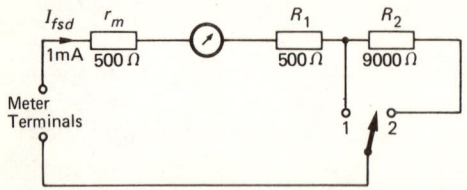

Fig.30.4

and a sensitivity of 1000 Ω/V.   The full-scale deflection current is then

$$\frac{1}{\text{Sensitivity}} = \frac{1}{1000} \text{ A} = 1 \text{ mA}$$

When the selector switch is in position 1,

Total resistance = r_m + R_1 = 500 + 500 = 1000Ω = 1 kΩ.
Total voltage or range = I_{fsd} × total resistance
= 1 mA × 1 kΩ = 1 V

When switch is in position 2,

Total resistance = r_m + R_1 + R_2 = 500 + 500 + 9000
= 10 000 Ω = 10 kΩ

Hence total voltage or range = 1 mA × 10 kΩ = 10 V.

Example (2)   Fig.30.5 shows a range selector network for an ammeter.   The basic meter has a resistance r_m = 810Ω and a full-scale deflection current = 1 mA.   When the

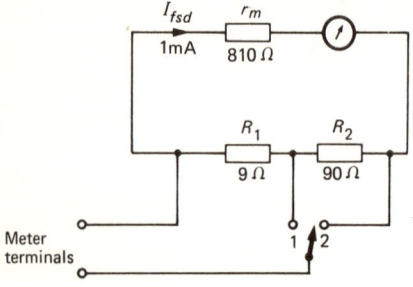

Fig.30.5

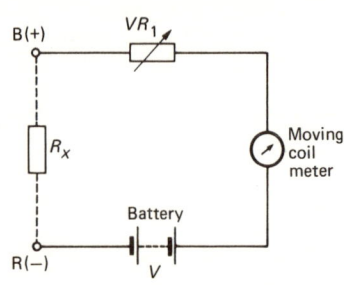

Fig.30.6 Basic ohmmeter

selector switch is in position 1, R_2 is connected in series with the basic meter giving a total resistance for that branch of

$$r_m + R_2 = 810 + 90 = 900\Omega$$

Resistor R_1 on the other hand is connected across the series branch.

Voltage across R_1 = voltage across series branch

$$= I_{fsd} \times (r_m + R_2)$$

$$= 1 \text{ mA} \times 900\Omega = 900 \text{ mV}$$

Current taken by shunt resistor $R_1 = \dfrac{900 \text{ mV}}{9\Omega} = 100 \text{ mA}$

Hence total current or range = 100 mA + 1 mA = 101 mA

When the switch is in position 2, both R_1 and R_2 are in parallel with the basic meter.

Total shunting resistance = $R_1 + R_2$ = 9 + 90 = 99Ω

Voltage across the shunt resistance = voltage across the basic meter

$$= V_{fsd} = I_{fsd} \times r_m = 1 \text{ mA} \times 810\Omega = 810 \text{ mV}$$

Current through the shunt resistance = $\dfrac{810 \text{ mV}}{99\Omega}$ = 8.07 mA

Hence total current or range = 8.07 mA + 1 mA = 9.07 mA.

The Ohmmeter

The ohmmeter is an instrument used for measuring d.c. resistance. There are two types of ohmmeters: the simple ohmmeter which is generally used in multimeters employing the principles of Ohm's law, and the MEGGER used to measure very high resistances such as cable insulations, etc. Only the first type will be discussed in some detail.

The SIMPLE OHMMETER consists of a moving coil meter, a battery and a variable resistor VR_1 as shown in Fig.30.6. Initially the terminals R and B are shorted. A current V/VR_1 flows through the meter; VR_1 is adjusted to give full-scale deflection which corresponds to zero resistance reading. This is known as the *zero setting* of the ohmmeter which should be carried out before the meter is used for measurements. When resistor R_x is then connected across terminals B and R, a current $V/(R_x + VR_1)$ flows giving a smaller deflection of the meter. The meter is calibrated in ohms so that direct resistance reading may be obtained.

The range of the ohmmeter may be *reduced* by connecting a resistance in series with the meter or *extended* by increasing the voltage of the battery.

As can be seen from Fig.30.6, the ohmmeter terminals have a specific polarity which must be observed when testing a diode, a transistor or an electrolytic capacitor.

The Multimeter

A multimeter is a moving coil instrument which incorporates a switching system enabling the meter to be used as an ammeter, a voltmeter, and an ohmmeter. Note that when the multimeter is switched to the ohmmeter range, the polarity of the meter terminals is reversed; the black terminal becomes positive while the red terminal becomes negative.

31 Components Testing

Testing of components is normally carried out by the use of an ohmmeter. The resistance of the component is measured and compared with that of a sound component.

Resistors and Inductors

The resistance of a resistor or an inductor may be measured accurately, thereby giving an indication of a possible fault. Resistors tend to increase their resistance or go open-circuit. They very seldom go short-circuit. Inductors (and transformers) go open-circuit which is easy to detect. They may also develop a shorted turn (i.e. a short circuit between turns) which is fairly common but difficult to detect by ohmmeter measurements, or may develop a short circuit between coil and core which is easy to confirm.

Capacitors

A capacitor may go open-circuit, short-circuit or leaky. A short-circuit or a leak (i.e. a low resistance reading) can easily be detected by ohmmeter measurement. On the other hand it is more difficult to confirm an open-circuit fault. When the ohmmeter terminals are connected across a capacitor, the ohmmeter battery voltage begins to charge up the capacitor. Provided the capacitor has a large value the meter may be observed to kick towards the zero reading and, as the capacitor charges up, its reading slowly rises to infinity. Failure to do that indicates an open-circuit. However, a small capacitor would charge up too quickly for the meter to register the movement.

Electrolytic capacitors may be tested for an open-circuit because of their high capacitance as well as their normally high leakage current which gives a low resistance reading (few 100 kΩ) when connected in the correct polarity.

Capacitors may change their value, in which case a bridge test where the value of the capacitance is measured is required to detect the fault.

Diodes

The testing of semiconductor devices normally involves the measurement of the forward and reverse resistance of a pn junction. For a diode the ohmmeter is first connected as shown in Fig.31.1a with a negative (red) terminal connected to the anode, and the positive (black) terminal

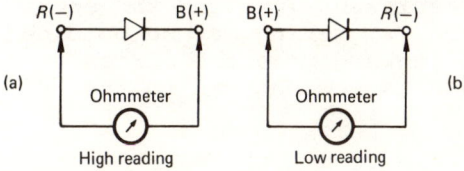

Fig.31.1 Testing a diode

connected to the cathode. The diode is, therefore, reverse biased giving a very high reading (in MΩ). The polarity is then reversed as shown in 31.1b giving a forward bias to the diode. A low forward resistance (600 - 1000Ω) will be registered on the meter. A diode giving a low reading (100Ω) in both directions may be assumed to be a short-circuit.

The ohmmeter test may also be used to identify the anode and cathode terminals of a diode. When the ohmmeter is reading low resistance (i.e. forward biased) as shown in 31.1b, the polarity of the ohmmeter is the same as the polarity of the diode, i.e. the positive (black) terminal is connected to the anode and the negative (red) terminal is connected to the cathode.

Bipolar Transistors

The bipolar transistor consists of two pn junctions which are tested individually in the same way as a diode junction. The forward and reverse resistance of the b-e and b-c junctions are measured separately. Normal readings are of the same order as for an ordinary diode. The resistance between the collector and the emitter is also tested and should give a very high reading (in MΩ) or infinity in both directions.

Junction FET

The ohmmeter is used to measure the resistance of both the junction as well as the channel path itself. For a sound fet the following readings should be obtained:

Drain-gate (forward bias) - low (40Ω).

Drain-gate (reverse bias) - infinity (MΩ).

Gate-source (forward bias) - low (40Ω).

Gate-source (reverse bias) - infinity (MΩ).

Drain-source or channel resistance (in either direction) - low (100Ω).

The SCR

Both the forward and reverse resistance between the anode and the cathode are very high. Triggering may be obtained by connecting the ohmmeter to forward bias the SCR as shown in Fig.31.2. If the gate is now shorted to the

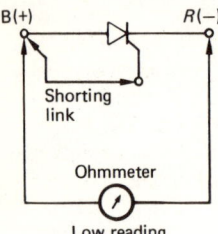

Fig.31.2 Testing an SCR

anode, gate current flows which triggers the SCR and makes it conduct. The ohmmeter reading thus drops to a low value (few 100Ω). The ohmmeter reading remains at that low value even if the shorting link is removed.

In-Circuit Component Testing

It is always advisable to check a suspected faulty component while it remains in the circuit. Once the fault is confirmed then a replacement may be made. These tests are carried with the aid of an ohmmeter and usually referred to as *continuity tests*. In carrying out in-circuit measurements, precautions have to be taken to ensure that the shunting effect of other components is minimized or taken into account.

IN-CIRCUIT MEASUREMENT OF RESISTANCE Consider the circuit in Fig.31.3 where R_x is the suspect component. An ohmmeter connected across R_x would read the total resistance

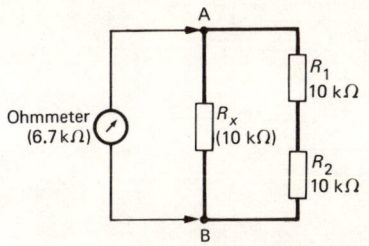

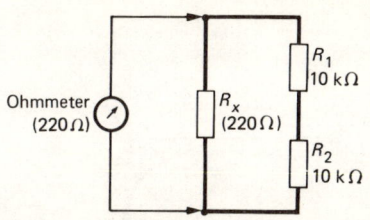

Fig.31.3 Shunting effect of
in-circuit measurement
of a resistance

Fig.31.4 Shunting effect of
in-circuit measurement
of a resistance

across A and B. Also $R_1 + R_2 = 10$ kΩ + 10 kΩ = 20 kΩ
offers a parallel path for the ohmmeter current to flow
giving a lower reading on the meter. If R_x is sound,
the ohmmeter reading would be

$$\frac{10 \times 20}{10 + 20} = \frac{200}{30} = 6.67 \text{ k}\Omega$$

If R_x is open-circuit, the ohmmeter would measure the
shunting resistance only, namely 20 kΩ.

The circuit in Fig.31.3 has a parallel path resistance
which is comparable to the resistance of the component
under test. Hence its effect has to be allowed for. In
Fig.31.4, $R_x = 220\Omega$. The shunting resistance is 10 kΩ +
10 kΩ = 20 kΩ which is 100 times greater than the component under test. In this case the effect of the shunting

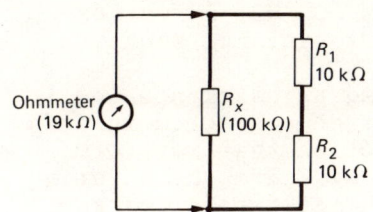

Fig.31.5 Shunting effect of
in-circuit measurement
of a resistance

path may be neglected.

In cases where the suspect component has a value far
greater than the total resistance of the shunting path
(Fig.31.5), an in-circuit test becomes unreliable. The
ohmmeter gives a reading approximately equal to the total
resistance of the shunting path. For the circuit in Fig.
31.5,

$$\text{Meter reading} = \frac{100 \times 20}{100 + 20} = \frac{2000}{120} = 19 \text{ k}\Omega$$

Note that a sound component always registers a reading
equal to or smaller (due to the shunting path) than its
real value and never a higher reading. If a higher
reading *is* registered, then the component has gone either
high or open-circuit.

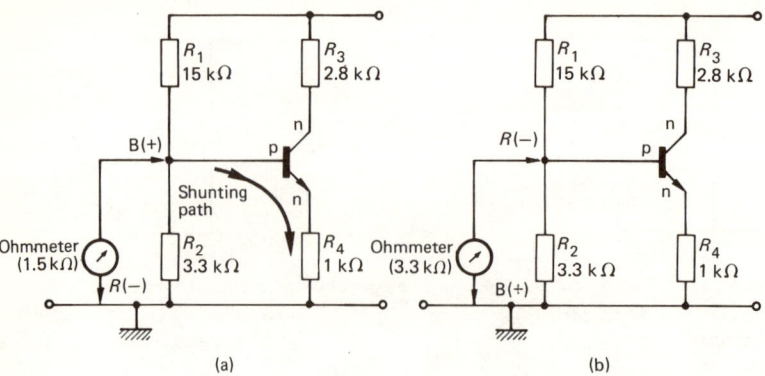

Fig.31.6 Shunting effect of in-circuit measurement of a resistance

In many cases a shunting path may include a pn junction either in the form of a diode or a b-e or a b-c junction of a transistor. Fig.31.6 shows a simple single stage transistor amplifier. To test resistor R_2 assume that the ohmmeter is connected with its positive terminal to the base and its negative terminal to chassis. The b-e junction becomes forward biased by the ohmmeter battery providing a low resistance shunting path through R_4 as shown in 31.6a and giving an unreliable reading. If on the other hand the terminals are reversed with the negative connected to the base and the positive to chassis as shown in 31.6b, the b-e junction becomes reverse biased. The shunting path now consists of a very high reverse resistance of the b-e junction (over 500 kΩ) in series with R_4. The effect of such a high resistance shunting path is of course negligible.

In general therefore when measuring a resistance in-circuit, the ohmmeter terminals must be connected in such a way as to reverse bias any pn junction that may form part of a parallel path in order to minimize the shunting effect. In practice two separate ohmmeter readings are taken with the ohmmeter terminals connected in both directions. *The higher reading is always the more accurate reading.*

INDUCTORS Inductors (or transformers) have very small resistances. Hence only a very low resistance shunting path has any noticeable effect on the accuracy of the ohmmeter reading.

CAPACITORS Leakage may be confirmed by in-circuit ohm-meter measurement test only if the resistance of the shunting path is calculated reasonably accurately. In testing an electrolytic capacitor care must be taken to ensure that the ohmmeter terminals are connected to provide the right polarity to the capacitor.

DIODES AND TRANSISTORS In testing a suspected diode or a transistor the forward and reverse resistances of the junctions are measured. In the forward direction the resistance of the junction is low, hence the effect of the shunting path is very small. For example, the forward resistance reading of the b-c junction of the transistor in Fig.31.6 is approximately 400Ω. In the reverse direction the resistance is very high with the ohmmeter reading mainly the resistance of the shunting path. For the reverse resistance of the b-c junction of the transistor in Fig.31.6, the ohmmeter reading will be approximately $R_3 + R_1$ = 3.3 kΩ + 15 kΩ = 17.8 kΩ.

Note that for an open circuit junction, both the forward and reverse readings will be the same (17.8 kΩ for an open circuit b-c junction in Fig.31.6).

32 The Effect of the Environment

Electronic equipment must be designed to work within certain environments. Temperature, humidity, altitude, radiation, dust and vibration are a number of considerations that have to be taken into account in the design as well as in the choice of passive and active components for any electronic piece of equipment.

Temperature change is the most important environmental consideration. Components that dissipate power such as a resistor or a transistor reduce their dissipation under high ambient (i.e. surrounding) temperature. Other components also suffer a change in value, poor stability, breakdown of insulation as temperature rises, or changes over a wide range.

A resistor rated at ½ W dissipates considerably less power as the ambient temperature increases beyond a specified maximum level. Under such conditions either the resistor is derated, i.e. made to operate at a lower power requirement, or a higher power rated resistor is used. Failure to do one or the other will increase the temperature of the resistor to a very high level resulting in a shorter life or damage to the component.

Valves do not suffer from high ambient temperatures since they have very high operating temperatures. Semiconductors on the other hand suffer in two ways.

(1) In the same way as a resistor, a semiconductor reduces its power dissipation with increasing ambient temperature. This is particularly important with power transistors where high power dissipation takes place. High ambient temperature thus limits the power output of the transistor. Dissipation may be increased by removing the heat away from the transistor with the aid of a HEAT SINK. A heat sink is a piece of metal which has good heat conductivity. The transistor is mounted on the heat sink to conduct the heat away. Large power transistors are made with a metal case that is directly connected to the collector. The metal case is then mounted on the heat sink (which may be the chassis itself) usually with a layer of silicon grease between the heat sink and the transistor. Silicon grease has a good heat conductivity, while at the same time it is a good electrical insulator.

(2) Semiconductor devices such as diodes and transistors change their characteristics as their temperature increases. Such changes may have to be taken into account and compensatory circuits used. Car radios for instance invariably employ a thermistor in the output stage to cancel out the change in the characteristics of the two output transistors as temperature rises.

33 Logical Fault Finding

Fault finding must follow a logical sequence, the purpose of which is to identify the cause of the fault and then to rectify it. The number of tests carried out should be kept to a minimum, and unnecessary or pointless tests must be avoided. Before any test is carried out, visual inspection of the circuit and the various components should be made to check for burnt-out components, obvious breaks in the printed circuit, and so on. Such inspection should take no more than two or three minutes and with experience will become instinctive. This having been done, without success, proper fault finding procedure can then begin.

Fault finding can be divided into two series of tests.

Series I known as DYNAMIC TESTS are applied to the complete electronic equipment in order to isolate the faulty stage. In some cases the fault symptoms point directly to the faulty stage or section. For example, a faulty TV receiver displaying a bright horizontal line across the screen (field collapse) indicates a fault in the field timebase.

Once the fault has been traced to a specific stage, then <u>Series II</u> or STATIC TESTS are applied to pin-point the cause of the fault to one or two possible faulty components such as a resistor, a capacitor, etc.

Series I Tests

These are the first set of tests that are carried out on the faulty electronic equipment. The tests must begin from the output going towards the input. Using the HALF-SPLIT METHOD the equipment is initially divided into two appropriate sections. The section at the output is tested first by INJECTING a signal similar to the one normally present at that point. If normal output is produced, the fault must then lie in the remaining section. This section is now divided into two appropriate sub-sections and the same procedure is repeated, and so on until the fault is isolated to the smallest distinguishable stage such as the output stage, i.f. or video amplifier, detector, and so on.

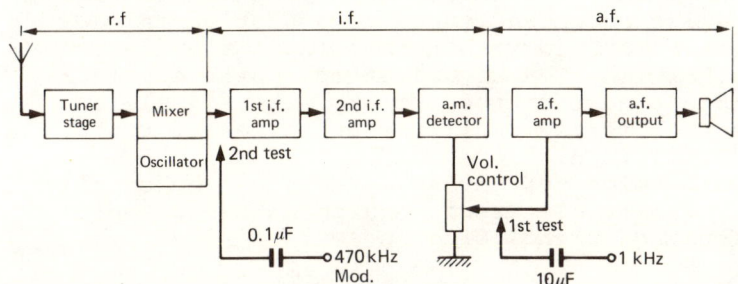

Fig.33.1

THE HALF-SPLIT METHOD APPLIED TO A RADIO RECEIVER
(Fig.33.1)
The most appropriate division is the a.f. on one hand and the i.f./r.f. on the other. The a.f. section is first tested by injecting a 1 kHz signal at its input (the volume control), via a coupling capacitor (10 - 50 µF). Low, distorted or no output indicates a faulty a.f. section. The a.f. section is now divided into two sub-sections: the output stage and the pre-amp. Each sub-section is then tested starting from the output. If the fault does not lie in the a.f. section, a pure 1 kHz tone should be heard from the loudspeaker. The fault must then be in the i.f./r.f. section.

A quick check on whether the fault is in the a.f. section or not is the SCREW-DRIVER test. Touch the input of the a.f. section (with the volume control on full) with a screw-driver. A noise from the loudspeaker should be heard if the a.f. section is sound.

If it is found that the fault lies in the i.f./r.f.
section a further division is made, namely an r.f. section and an i.f. section. The i.f. section is tested
first by injecting an amplitude-modulated 470 kHz signal
via a coupling capacitor (0.01 - 0.1 µF) into the input
of the i.f. section, e.g. into the base of the first i.f.
transistor. For f.m. receivers a frequency-modulated
10.7 MHz test signal should be used. A sound i.f.
section will produce a pure tone (400 - 600 Hz) at the
loudspeaker. Otherwise further sub-divisions are made
until the faulty stage is discovered, e.g. one i.f.
amplifier or the detector.

If the fault is in the r.f. section, this section is
sub-divided if possible and tested as follows. An
amplitude-modulated 1000 kHz signal is injected via a
coupling capacitor (0.01 - 0.1 µF) into the input of the
stage. The receiver should then be tuned to receive the
radio frequency of 1000 kHz or 300 m on the MW. Another
appropriate frequency must be used for f.m. receivers.

An alternative to signal injection is the SIGNAL
TRACING method where a radio receiver or other electronic
equipment is switched on and tuned to a station in the
case of a receiver or fed with normal input signal. The
presence or otherwise of a signal at appropriate points
is confirmed starting from the output. The signal is
observed on a cathode ray oscilloscope to test for its
waveshape and amplitude.

The above principles for Series I tests can be equally
applied to any faulty electronic equipment using appropriate divisions and signal frequencies.

Principles of Series II Tests

Having identified the faulty stage, this series of tests
is used to identify the faulty component.
(1) Start with d.c. tests. Use a meter having a
    sensitivity of at least 20 000 Ω/V.
(2) Take voltage measurements only. If current value is
    required, estimate it by measuring the voltage across
    a known resistor.
(3) For confirming a suspected component, apply in-circuit
    measurement techniques using an ohmmeter (see section
    30). When ohmmeter tests are made d.c. supply voltage
    must be switched off.
(4) If the fault is not revealed by d.c. measurements,
    then, and only then, apply signal tests to the stage.

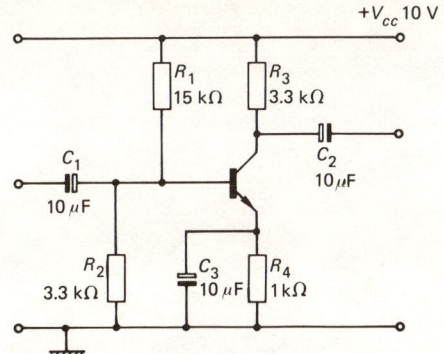

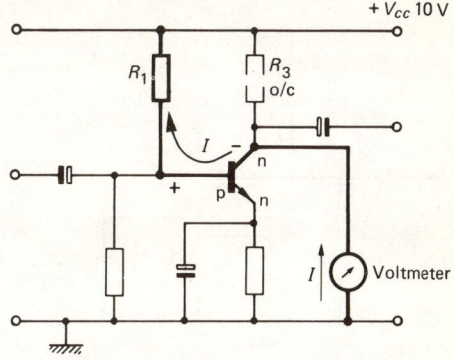

Fig.33.2 Normal readings:
e 1.1 V, b 1.72 V, c 6.37 V

Fig.33.3 R_3 o/c, transistor cut
off: e 0.3 V, b 0.94 V, c 0.3 V

APPLICATION TO SINGLE STAGE AMPLIFIER (refer to Fig.33.2)
Usually the normal d.c. readings for the stage are given.
If not they could be estimated to a reasonable degree of
accuracy. The voltage readings of the faulty stage are
then compared with the normal readings. This comparison
should then give a good indication of where the fault lies.
First determine the static state of the transistor:
(*a*) transistor at cut-off producing no output signal (or
     going towards cut-off) or
(*b*) transistor at saturation producing low distorted
     output or going towards saturation or
(*c*) transistor at normal static state.
Once the state of the transistor has been established,
the next step is to work out the cause for cut-off or
saturation. (When functioning under normal static state,
the fault is an a.c. fault which will be dealt with later.)

CUT-OFF CONDITION
Transistor current ceases to flow, i.e. cut-off, if (*a*) b-e
forward bias is zero or (*b*) the current path is interrup-
ted, namely when R_3 is open circuit (o/c), when R_4 is o/c,
or the transistor itself is faulty. Normally when the
transistor is at cut-off, the collector goes up to d.c.
supply V_{cc}. However, if R_3 goes o/c the collector is
"floating" and theoretically it is at base potential. As
soon as a voltmeter is connected to measure the collector
voltage, the base-collector junction becomes forward bias
as shown in Fig.33.3. Current will flow through the meter,
the b-c junction and R_1 giving a small reading on the
meter. This reading is entirely due to the meter resistance.
   Similarly, when cut-off is caused by R_4 going o/c, the
emitter is "floating" and theoretically should be at base

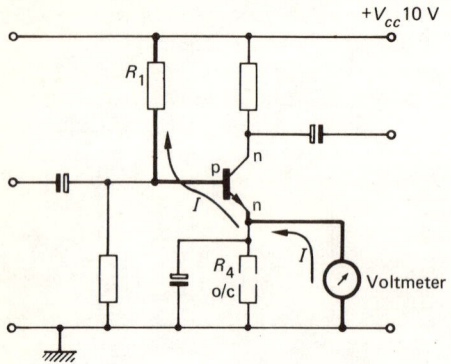

*Fig.33.4 R_4 o/c, transistor cut
off: e 1.25 V, b 1.74 V, c 10 V*

potential. However, a voltmeter connected to measure
that voltage provides a path for current to flow and
forward biases the base-emitter junction, giving a reading on the meter slightly higher than normal (Fig.33.4).
    Table 33.1 gives a summary of the above faults. Note
that in the table a high V_{be} refers to an increase of
between 0.1 to 0.2 V in the base-emitter forward bias.

TABLE 33.1 Cut-off Condition

FAULT		CAUSE
(1) e	zero	
b	zero	
c	V_{cc}	R_1 o/c
V_{be}	zero	
(2) e	high	
b	normal	
c	V_{cc}	R_4 o/c
V_{be}	low	
(3) e	low ($\simeq V_c$)	
b	low	
c	low ($\simeq V_e$)	R_3 o/c
V_{be}	normal	

TRANSISTOR FAULTS also produce cut-off conditions. Voltage readings in such cases depend on the nature of the fault as well as the component values. For example, a b-e s/c (Fig.33.5) cuts the transistor off and puts R_4 in parallel with R_2 causing base and emitter voltage to go down to a low value determined by the potential divider R_1-R_2/R_4. Collector voltage will, of course, be at V_{cc}. Fig.33.6 shows the effect of a s/c between emitter and collector.

Other transistor faults are given in Table 33.2.

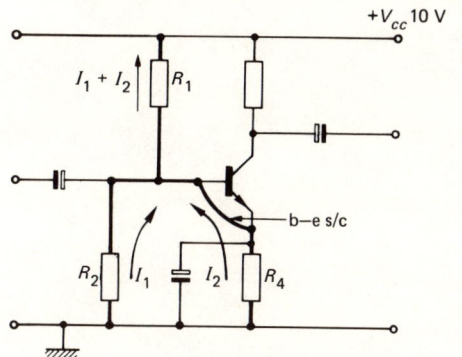

Fig.33.5 *b-e s/c, transistor cut off: e 0.48 V, b 0.48 V, c 10 V*

Fig.33.6 *e-c s/c, transistor cut off: e 2.29 V, b 1.77 V, c 2.29 V*

TABLE 33.2

FAULT		CAUSE
(1) e	zero	
b	normal	
c	V_{cc}	b-e o/c
V_{be}	very high, cannot be sustained by functioning junction	
(2) e	low	
b	low	
c	V_{cc}	b-c o/c
V_{be}	normal	

SATURATION CONDITION
As explained in section 10, transistor current is determined by the forward bias across the base-emitter junction.
A small increase in this bias produces a large rise in
transistor current. When the current through the
transistor is at maximum value, the transistor is said to
be saturated or bottomed. Collector voltage falls as
current increases until at saturation it is almost the
same as the emitter voltage to within 0.1 - 0.5 V. In
general at saturation, the emitter, base and collector
potentials are approximately at the same level. Table
33.3 summarises.

TABLE 33.3

FAULT			CAUSE
(1)	e	high (V_c)	R_2 o/c
	b	high	or
	c	low	R_1 low
(2)	e	zero	
	b	low	C_e s/c
	c	very low	

NORMAL STATIC STATE
Normal d.c. voltages with no or low output indicate an
a.c. fault, e.g. an open circuit coupling capacitor.
Before replacing a suspected o/c capacitor, place a good
capacitor of similar value across the suspected component
to confirm the fault. An open circuit decoupling
capacitor (C_3 in this case) produces low undistorted
output. A leaky or short circuit coupling capacitor
usually introduces a change in the d.c. state of the
transistor depending on the preceding or following stages.

In fault finding the following should be remembered:
(1) Do not jump to conclusions by merely comparing one
    reading with its normal value. The whole set of
    readings (in the case of a transistor stage the
    emitter, base and collector) should be noted on
    paper and compared with the normal set of readings.
(2) If accurate readings (up to 0.01 V are possible on
    a 20 000 Ω/V meter) are taken, then in the vast
    majority of cases two equal readings indicate a s/c.
    However, this is not always true and further checks
    must be made before the final decision is made.

Fault Finding Examples

In each example separate fault conditions are described. The fault in each case is assumed to be the result of a failure in a single component, but there may be more than one component that could cause the fault. All test potentials are measured in volts with respect to the chassis, using a 20 000 Ω/V voltmeter.

(1) D.C. Coupled Amplifier
(Refer to Fig.33.7)

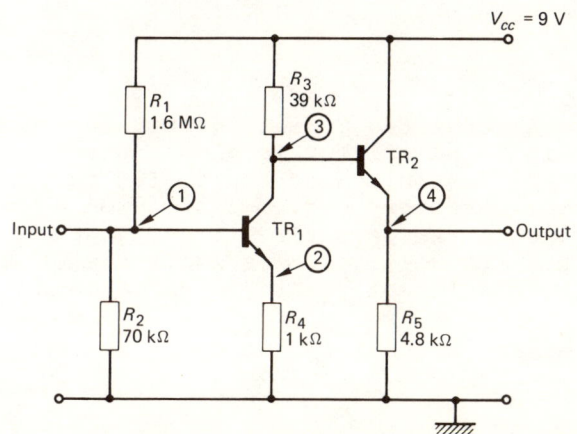

Fig.33.7

Test point	1	2	3	4
Normal readings	0.7	0.1	5.1	4.5
Fault A (no output)	0	0	8.65	8
Fault B (no output)	0.01	0.01	8.4	7.75
Fault C (no output)	0.48	0	0.02	0
Fault D (no output)	0.7	0.1	5.1	4.7

FAULT A   TR$_1$ is cut off, emitter is zero, and collector is at V_{cc}. Cut-off caused by TR$_1$ base (test point 1) having zero volts. Since the two transistors are d.c. coupled, TR$_2$ base voltage increases with collector of TR$_1$, causing TR$_2$ to saturate.

   Answer: R$_1$ o/c

FAULT B TR_1 is cut off. Voltages at test points 1 and 2 are equal suggesting a short circuit. This voltage is determined by resistor chain R_1 (1.6 MΩ) in series with R_2 (70 kΩ) with the latter shunted by R_4 (1 kΩ). As in Fault A, TR_2 is saturated.

Answer: TR_1 b-e s/c.

FAULT C TR_1 and TR_2 are both cut off. TR_2 emitter voltage is at zero (test point 4) indicating cut off. TR_1 has low collector voltage suggesting saturation, while its emitter (test point 2) is zero. Hence TR_1 must be at cut off caused by the collector resistor going open circuit. TR_1 base is down because of a large base current flowing through high resistor R_1 (1.6 MΩ).

Answer: R_3 o/c.

FAULT D TR_1 is normal. A saturated TR_2 would produce a high emitter voltage. Since the emitter is almost normal and with the absence of any possible a.c. fault, TR_2 must be at cut off. A faulty transistor, e.g. b-c o/c, would produce zero emitter voltage. An open circuit b-e junction would produce low emitter voltage. The fault must be R_5 o/c with emitter reading (test point 5) due to meter resistance.

Answer: R_5 o/c.

(2) Two-stage R-C Coupled FET Amplifer

(Refer to Fig.33.8)

Test point	1	2	3	4	5	6
Normal readings	15	2.4	5.55	0	3.3	12.75
Fault A (no output)	15	0	0	0	3.3	12.75
Fault B (no output)	15	2.4	5.55	5.55	7.5	9.9
Fault C (no output)	15	2.4	5.55	0	4.1	15
Fault D (no output)	15	2.4	5.55	0	6.1	10.8
Fault E (low output)	15	2.4	5.55	0	3.3	12.75

FAULT A VT_1 cut off. VT_2 normal. Zero drain and source voltages suggests R_2 o/c. Note that if R_3 were o/c, a high source reading would have been obtained.

Answer: R_2 o/c.

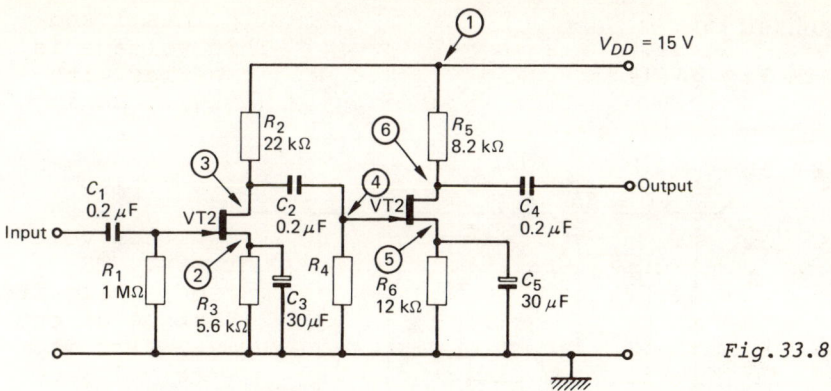

Fig.33.8

FAULT B VT$_1$ normal. VT$_2$ saturated (source high, drain low). A fet may saturate if the gate-source reverse bias V_{GS} is reduced. With an n-channel fet used in the circuit, this may be produced either by the gate voltage going positive (towards the source) reducing V_{GS} or by removing the gate leakage resistor (also see Fault D). Test points 3 and 4 are equal suggesting C_2 short circuit. This will increase VT$_2$ gate voltage which reduces the reverse bias and increases the current through the transistor.

    Answer: C_2 s/c.

FAULT C VT$_1$ normal. VT$_2$ drain at 15 V suggests cut-off. The increase in source voltage (test point 5) points to R_6 o/c with the voltage reading at the source due to meter resistance. Note that a faulty fet such as an open circuit junction would produce cut-off but would give a zero source reading.

    Answer: R_6 o/c.

FAULT D VT$_1$ is normal. VT$_2$ is saturated (high source, low drain). Since gate voltage remains at zero, saturation is due to gate leakage resistor going open circuit, thus removing the gate-source reverse bias and increasing the drain current.

    Answer: R_4 o/c.

FAULT E All d.c. voltages are normal. Hence fault is a.c. Low output suggests an open circuit decoupling capacitor introducing negative feedback and reducing the gain of the amplifier.

    Answer: C_3 or C_5 o/c.

(3) Stabilised Power Supply

(Refer to Fig.33.9)

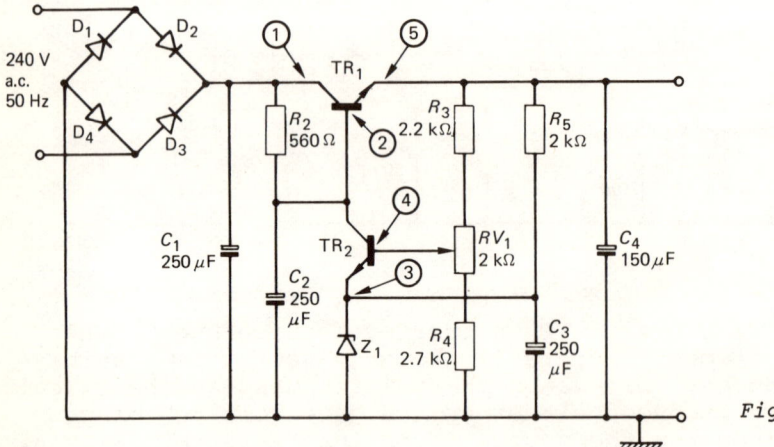

Fig.33.9

Test point	1	2	3	4	5
Normal readings	18	14.1	8.75	9.35	13.3
Fault A	19.5	2.65	0	0.7	2.0
Fault B	19.8	13.5	8.75	9.35	13.5
Fault C	20	20	8.65	0	19.3
Fault D	19.5	0	0	0	0
Fault E	20	20	0	0	0

Note Always start by noting the zener reference voltage. A low or zero reading indicates that the zener is not at breakdown either because of a faulty zener or a fault in the resistor feeding it (R_5 in Fig.33.9).

FAULT A   Zener reference voltage is zero, which points to
a short circuit between test point 3 and chassis.   Note
that an open circuit R_5 would not give zero reading at
test point 3 since the meter resistance would complete
the emitter circuit of TR_2 giving a voltage reading.

   Answer: Z_1 or C_3 s/c.

FAULT B   Zener voltage is normal.   Test points 2 and 5
are the same potential suggesting a short circuit. TR_1
is cut-off, hence its collector (test point 1) has
increased.

   Answer: TR_1 b-e s/c.

FAULT C   Zener voltage is 0.1 V lower than normal indicating that, while the zener is at breakdown, it is taking
a smaller current then normal.   This is due to TR_2 being
at cut-off (note its collector at 20 V).   The cut-off is
due to zero volts at TR_2 base indicating a break in the
bias chain.

   Answer: R_3 or upper part of RV_1 o/c.

FAULT D   The zero volt at the zener is caused by the zero
volts at test point 5. TR_2 is at cut-off (base and emitter
are both zero volts).   Test point 1 is almost normal yet
TR_2 collector is at zero volts.   This indicates a fault
in components R_2 or C_2.

   Answer: R_2 o/c or C_2 s/c.   (C_2 s/c would anyway cause
R_2 to go o/c due to the large current through it.)

FAULT E   Zero zener voltage is caused by zero voltage at
test point 5.   Voltage across b-e junction of TR_1 = test
point 2 - test point 5 = 20 - 0 = 20 V.   A pn junction
cannot sustain a forward bias higher than 0.7 V.

   Answer: TR_1 b-e o/c.

(4) Output Stage

(Refer to Fig.33.10)

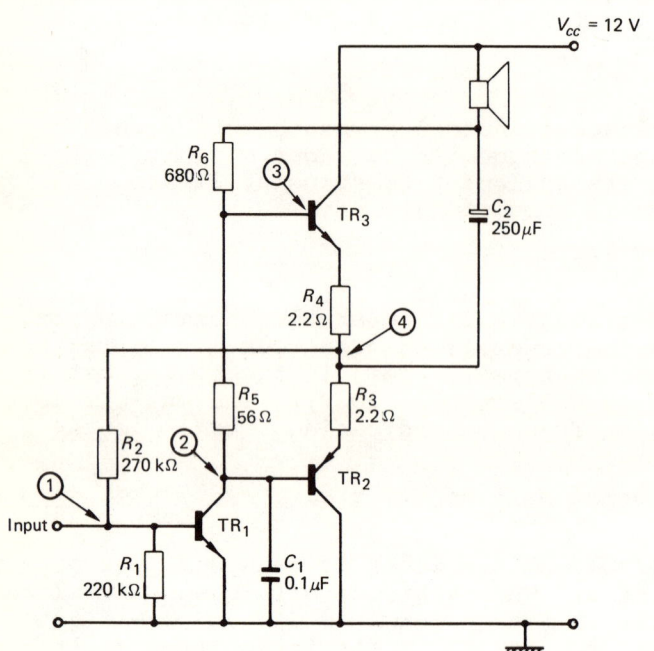

Fig.33.10

Test point	1	2	3	4	5
Normal readings	0.61	6.3	6.7	6.45	12
Fault A (no output)	0	12	12	12	12
Fault B (no output)	0.19	0	0.79	1.09	12
Fault C (very low output)	0.4	1.1	1.08	1.08	12
Fault D (no output)	0.61	5.85	6.22	6.0	12

Note Start by looking at test point 4 where the normal reading should be approximately $\frac{1}{2}V_{cc}$. If test point 4 is high, then TR_3 is conducting more than TR_2 and vice versa.

FAULT A Test point 4 is at V_{cc}. TR_3 is saturated and TR_2 is cut off. TR_1 collector (test point 2) is at V_{cc} indicating TR_1 cut off. TR_1 base is zero volts, hence fault is in bias chain.

Answer: R_2 o/c.

FAULT B Test point 2 is zero volts suggesting a short circuit to chassis. Such a short circuit would cause TR_2 b-e junction to conduct heavily, charging C_2 and reducing the potential at test point 4. Low TR_1 base voltage is due to low voltage at test point 4.

Answer: TR_2 b-c s/c or C_1 s/c.

FAULT C Test point 4 is low. C_2 is charged by TR_2 which is conducting more than normal due to its low base voltage (test point 2). The low potential at test point 2 is not due to a saturated TR_1. Note its low b-e forward bias V_{be} of 0.4 V instead of the normal 0.61 V. TR_1 must thus be at cut-off caused by open circuit collector resistance. TR_1 is not fully cut-off. It conducts through the b-e junction of TR_2 thus charging C_2 and bringing test point 4 down to 1.08 V.

TR_3 base is at the same potential as TR_1 collector since no current flows through R_5. TR_3 is at cut-off.

Answer: R_6 o/c.

FAULT D Readings are approximately normal. Hence the fault is a.c. An open circuit C_2 would produce a voltage at test point 4 slightly lower than normal, since the balance between the two output transistors TR_2 and TR_3 is no longer produced by the charge on capacitor C_2.

Answer: C_2 o/c.

(5) Bistable Multivibrator

(Refer to Fig.33.11). Normal readings are as follows.

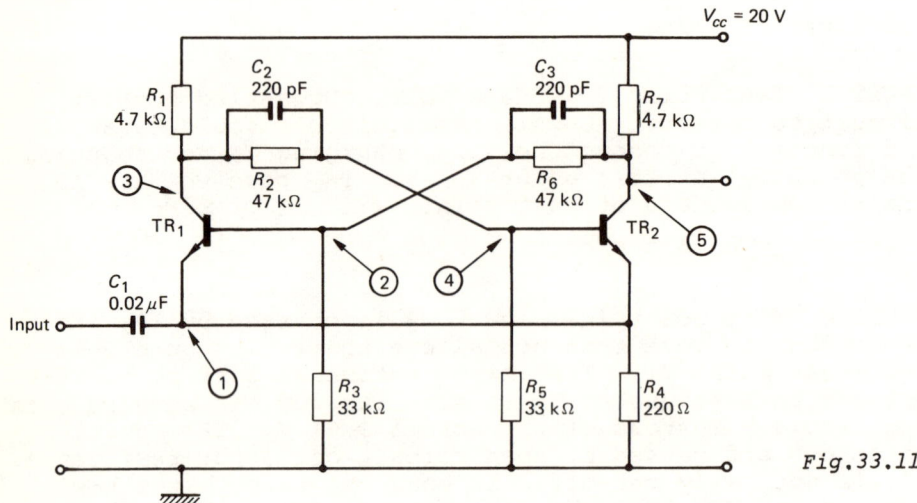

Fig.33.11

Test point	1	2	3	4	5
State I (TR$_1$ off)	1.1	0.45	18.5	1.3	1.2
State II (TR$_1$ on)	1.1	1.3	1.2	0.45	18.5

In each of the following cases the output fails to change when an input pulse is received.

Test point	1	2	3	4	5
Fault A	0.95	0.95	18.5	1.6	1.0
Fault B	0.95	0	18.5	1.6	1.0
Fault C	0.1	0.7	0.1	0.05	18.5
Fault D	1.1	0.45	18.5	1.3	1.2
Fault E	7.7	8.0	19.0	8.0	19.0

Note The bistable has two normal states in which it may settle, hence two normal sets of readings. When locating a fault, first determine the state of each transistor, then compare the faulty readings with the normal readings to determine the reason for the failure to change over from one state to another.

FAULT A TR_1 off, TR_2 on. TR_1 b-e voltage is zero suggesting a short circuit between base and emitter. Voltage at test point 1 (0.95 V) is now reduced due to the shunting effect of R_4 upon R_3.

Answer: TR_1 b-e s/c.

FAULT B TR_1 off, TR_2 on. TR_1 base voltage is zero suggesting a faulty bias chain R_6-R_3.

Answer: R_6 o/c.

FAULT C TR_2 is cut off. TR_1 seems saturated, however the low emitter voltage (test point 1 at 0.1 V) indicates cut-off due to open circuit collector resistor R_1. TR_1 collector voltage reading (test point 3) is due to meter resistance. Emitter voltage reading (test point 1) is due to a large base current flowing through forward biased b-e junction of TR_1.

Answer: R_1 o/c.

FAULT D D.C. readings are normal. Hence fault is a.c.

Answer: C_1 o/c.

FAULT E The readings are symmetrical with both TR_1 and TR_2 at cut-off. Hence the fault must be common to both transistors. R_4 is the only common component that affects the d.c. conditions of the circuit. Reading at test point 1 is due to meter resistance.

Answer: R_4 o/c.

(6) Fault Finding Using an Ohmmeter

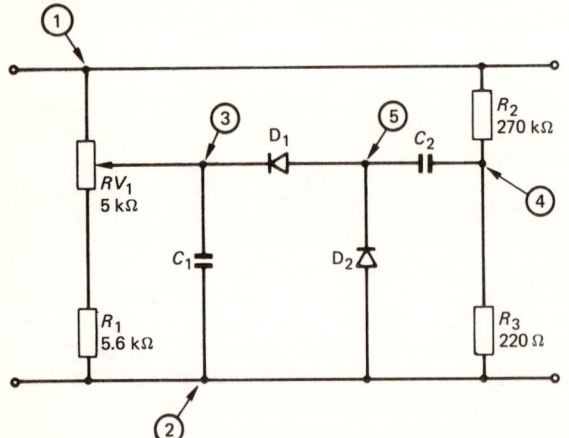

Fig.33.12

(Refer to Fig.33.12). The polarity signs given in brackets indicate the polarity of the ohmmeter leads connected to that particular test point. The slider on the potentiometer RV_1 is set to its mid-position.

FAULT A   1(+) and 2(−) ≃ 10 kΩ
                   1(−) and 2(+) ≃ 10 kΩ
                   3(+) and 2(−) ≃ 8.1 kΩ
                   3(−) and 2(+) ≃ 8.1 kΩ
                   5(+) and 2(−) ≃ 8.9 kΩ
                   5(−) and 2(+) = ∞

FAULT B   1(+) and 2(−) ≃ 10 kΩ
                   1(−) and 2(+) ≃ 2.8 kΩ
                   4(+) and 2(−) ≃ 220 Ω
                   4(−) and 2(+) ≃ 180 Ω

FAULT C   1(+) and 2(−) ≃ 270 kΩ
                   1(−) and 2(+) ≃ 3.9 kΩ
                   3(+) and 2(−) ≃ 273 kΩ
                   3(−) and 2(+) ≃ 1.4 kΩ
                   4(+) and 1(−) ≃ 4 kΩ
                   4(−) and 1(+) ≃ 270 kΩ

Note Take each reading in turn and compare it with a roughly estimated normal resistance taking account of shunting paths, forward and reverse pn junctions, etc. By a process of elimination one component will be found faulty.

FAULT A With ohmmeter connected to 1(+) and 2(-), D_1 and D_2 are reverse biased and a reading of 10 kΩ is normal. Reversing the connection, D_1 and D_2 become forward biased. A lower resistance reading would be expected. Since in both cases the ohmmeter gives the same reading, one of the diodes must be open circuit. When the ohmmeter is connected across 5(+) and 2(-), it forward biases D_1 and reverse biases D_2 and gives a normal reading. Reversing the connections the ohmmeter should register a resistance of approx. 700 Ω due to forward biased D_2. An infinite reading indicates a faulty D_2.

Answer: D_2 o/c.

FAULT B Reading across 1(+) and 2(-) is normal. Reversing the connection, forward biases both diodes. The normal reading is approx. 3.6 kΩ. However the reading on the ohmmeter is lower (2.8 kΩ) indicating the introduction of a shunting resistance when one or both diodes are forward biased. Reading across 4(+) and 2(-) gives a normal reading of 220 Ω since C_2 is an open circuit. Reversing these connections should give the same reading since C_2 remains o/c. A lower reading indicates a short circuited C_2 in which case D_1 is forward biased introducing a shunt across R_3.

Answer: C_2 s/c.

FAULT C Reading across 1(+) and 2(-) is very high indicating a fault in the main path RV_1-R_1 with the meter reading the high resistance of the shunting path R_2-R_3. With the ohmmeter reversed, D_1 and D_2 are forward biased giving a reading only slightly higher than normal. This confirms that the fault is before D_1. Reading across 3(+) and 2(-) is high (273 kΩ) indicating that the top part of RV_1 is sound. Hence lower half of RV_1 or R_1 is o/c. Remaining tests only confirm this conclusion.

Answer: Lower half of RV_1 o/c or R_1 o/c.

A1 Types of Resistor

TYPE	POWER RATING	STABILITY	TOLERANCE	OTHER POINTS
Wire-wound	High (up to 300 W)	High (1%)	5%	Large size Expensive
Carbon	Low (up to 3 W)	Low (20%)	5–20%	Cheap Small size Widely used
Metal or oxide film (precision)	Low (less than 2 W)	High (1%)	Very close	Expensive Small size
Metal or oxide film (power)	High (up to 300 W)	High (1%)	Close	Large size Expensive

A2 Types of Capacitor

TYPE	CAPACITANCE RANGE	FREQUENCY RANGE	OTHER POINTS
Paper	0.001 – 1 µF	Medium (50 Hz – 1 MHz)	Cheap
Polystyrene	0.001 – 1 µF	Medium	Medium stability (1 – 5%)
Ceramic	up to few µF	v.h.f. and u.h.f.	Small size High stability
Mica and silver mica	a few pF to 0.05 µF	r.f. (200 MHz or over)	Small size Robust High stability Expensive
Air	very low (up to 0.0005 µF)	v.h.f. and u.h.f.	Bulky Widely used as variable tuning capacitors
Electrolytic	high (up to 1000s µF)	Low (d.c. to a.f.)	Cheap Very high tolerance (over 50%) Widely used

A3 List of SI Units

Angle	radians rad
Angular velocity	rad/s
Capacitance	farad F
Charge	coulomb C
Current	ampere A
Electromotive force or electric potential	volt V
Energy	joule J
Force	newton N
Frequency	hertz Hz
Impedance	ohm Ω
Inductance	henry H
Length or distance	metre m
Magnetic flux	weber Wb
Magnetic flux density	tesla T
Mass	kilogramme kg
Power	watt W
Pressure	N/m^2
Reactance	ohm Ω
Resistance	ohm Ω
Time	second s
Torque	newton-metre Nm
Velocity or speed	m/s
Weight	newton N
Work	joule J

A4 General Technology: Multiple Choice Question Paper

(1) The reverse resistance of a germanium diode is 2 MΩ at an ambient temperature of 25°C. If the ambient temperature is raised to 70°C the reverse resistance will be approximately

(a) 1 kΩ
(b) 100 kΩ
(c) 2 MΩ
(d) 5 MΩ

(2) Refer to Fig.Q1. If D_1 is a silicon device the expected voltage between points A and B will be

(a) 0 V
(b) 0.3 V
(c) 0.6 V
(d) 5 V

(3) Refer to Fig.Q1. The voltage across R_1 is

(a) 0 V
(b) 0.6 V
(c) 5 V
(d) 10 V

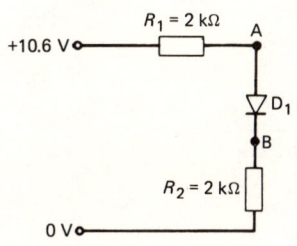

Fig.Q1

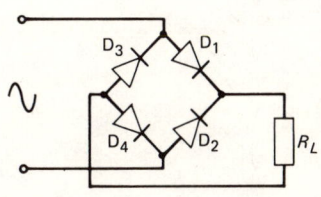

Fig.Q2

(4) Refer to Fig.Q2. With a mains input of 240 V r.m.s. the peak inverse voltage across each diode will be approximately

(a) 170 V
(b) 240 V
(c) 350 V
(d) 700 V

(5) The purpose of the suppressor grid in a pentode valve is to

(a) improve the amplification
(b) reduce the secondary emission effect
(c) increase the secondary emission effect
(d) isolate the anode

(6) Refer to Fig.Q3.   The current taken by the zener is

(a) 20 mA
(b) 18 mA
(c) 2 mA
(d) 0

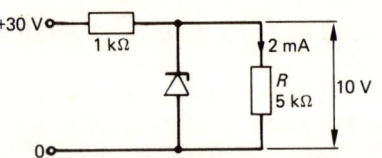

Fig.Q3

(7) Refer to Fig.Q3.   If the resistance of the load resistor R is increased to 10 kΩ, the current in the zener diode will

(a) decrease by 1 mA
(b) remain constant
(c) increase by 1 mA
(d) increase by 2.0 mA

(8) The anode current of an SCR can be cut off by applying a

(a) zero voltage to the gate terminal
(b) negative voltage to the gate terminal
(c) negative voltage to the cathode terminal
(d) negative voltage to the anode terminal

(9) A transmission line has a loss of 6 dB.   If a signal of 1 V is fed in, the voltage appearing at the other end will be

(a) $\frac{1}{6}$ V

(b) $\frac{1}{3}$ V

(c) $\frac{1}{2}$ V

(d) 1 V

(10) Refer to Fig.Q4.   In the basic phase-splitter circuit shown, the ratio of R_1 to R_2 must be

(a) 1 : 2
(b) 1 : 1
(c) 2 : 1
(d) 6 : 1

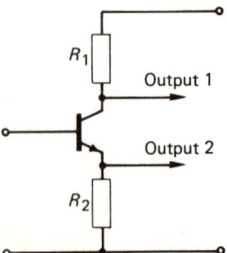

Fig.Q4

(11) If a resistance is shunted across the tuned load of a radio frequency amplifier, the effect will be to increase the

    (a) gain of the amplifier
    (b) bandwidth of the amplifier
    (c) selectivity of the amplifier
    (d) tuned frequency of the amplifier

(12) The low frequency response of an amplifier is greatly improved if the coupling is by

    (a) resistor and capacitor
    (b) transformer
    (c) choke and capacitor
    (d) direct connection

(13) If a typical domestic receiver is tuned to 1100 kHz and has an i.f. of 470 kHz, the frequency of the local oscillator is

    (a) 470 kHz
    (b) 1100 kHz
    (c) 1570 kHz
    (d) 10 MHz

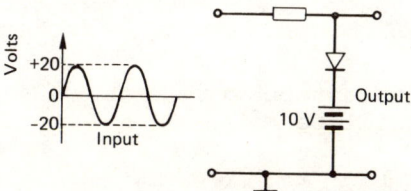

Fig.Q5

(14) Refer to Fig.Q5. Assuming an ideal diode, the output waveform produced by the circuit shown will be

a

b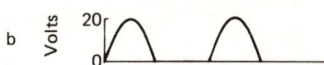

c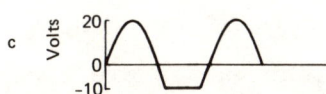

d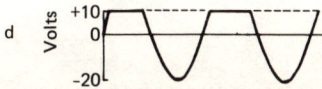

(15) Refer to Fig.Q6. The binary counter has been re-set to zero. After eleven input pulses, the output states will be

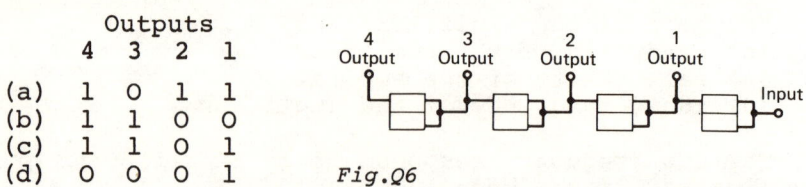

	Outputs			
	4	3	2	1
(a)	1	0	1	1
(b)	1	1	0	0
(c)	1	1	0	1
(d)	0	0	0	1

Fig.Q6

(16) A binary counter which uses bistable elements only is required to count 12 pulses. The minimum number of bistables required to do this is

(a) 3
(b) 4
(c) 6
(d) 12

(17) If the area of the plates of a capacitor and the distance between them are both doubled the capacitance will be

(a) reduced by $\frac{1}{3}$
(b) unaltered
(c) tripled
(d) increased by 6 times

(18) Refer to Fig.Q7. The instanteous value of the voltage after $\frac{1}{4}$ cycle will be

(a) 0 V
(b) + 50 V
(c) + 70 V
(d) + 100 V

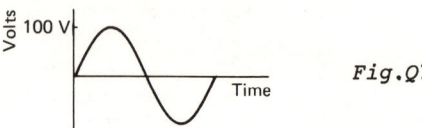

Fig.Q7

(19) If a car is travelling at 2 m/s and reaches a speed of 12 m/s after 2 s, its acceleration is

(a) 0.5 m/s^2
(b) 2 m/s^2
(c) 5 m/s^2
(d) 12 m/s^2

(20) If three capacitors of different values are connected in series across a 60 V d.c. supply, the voltage across each capacitor will be

(a) 20 V
(b) 60 V
(c) inversely proportional to its capacitance
(d) directly proportional to its capacitance

(21) The heat produced per minute from a 1000 W electric fire is

 (a) 2 J
 (b) 1000 J
 (c) 60 000 J
 (d) 250 000 J

(22) Refer to Fig.Q8. The oscillator used is known as

 (a) Hartley
 (b) Tuned Collector
 (c) Tuned base
 (d) Blocking

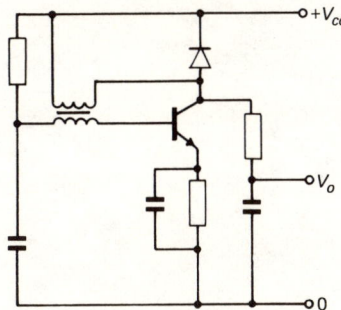

Fig.Q8

(23) Refer to Fig.Q8. The output waveform from the circuit shown is

 (a) sinusoidal
 (b) sawtooth
 (c) square
 (d) pulse

(24) When tested, an electrolytic capacitor is found to have no leakage current. This indicates that the capacitor

 (a) is s/c
 (b) is in good condition
 (c) is o/c
 (d) has a low d.c. working voltage

(25) Refer to Fig.Q9. If the input is 10 V d.c., the output voltage is

 (a) 0 V d.c.
 (b) 1 V d.c.
 (c) 9 V d.c.
 (d) 10 V d.c.

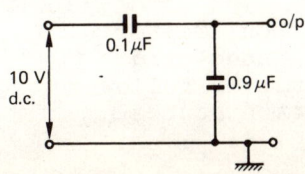

Fig.Q9

(26) The illumination of a surface at a distance of 2 m from a point source of light is 10 lux. What is the illumination if the distance is decreased to 1 m?

   (a) 5 ℓ
   (b) 10 ℓ
   (c) 20 ℓ
   (d) 40 ℓ

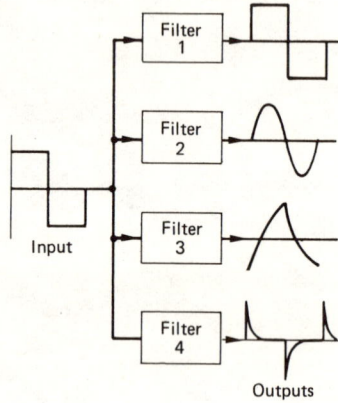

Fig.Q10

(27) Refer to Fig.Q10. A square waveform is applied to the inputs of four low-pass filters. Which filter has the highest cut-off frequency?

   (a) 1
   (b) 2
   (c) 3
   (d) 4

(28) If the readings obtained on a power supply are 100 V on no-load and 90 V on full-load, the regulation of the power supply is

   (a) 5%
   (b) 9%
   (c) 10%
   (d) 11%

(29) Which circuit connection would be used to match a high impedance circuit to a low impedance load?

   (a) A common emitter
   (b) A grounded grid
   (c) An emitter follower
   (d) A common source

(30) The low frequency response of a fet audio amplifier is limited by the

  (a) coupling capacitor
  (b) leakage current
  (c) inter-electrode capacitances
  (d) value of load resistor

(31) Refer to Fig.Q11. In the circuit shown, the minimum power rating of resistor R_1 will be

  (a) $\frac{1}{8}$ W

  (b) $\frac{1}{4}$ W

  (c) $\frac{1}{2}$ W

  (d) 1 W

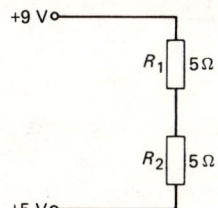

Fig.Q11

(32) Refer to Fig.Q12. The base voltage is

  (a) - 10 V
  (b) - 9.9 V
  (c) - 5 V
  (d) - 0.1 V

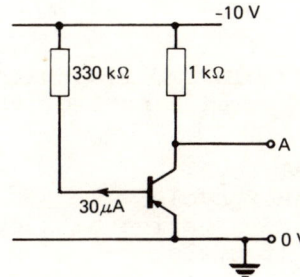

Fig.Q12

(33) Refer to Fig.Q12. The voltage at A is - 7 V. The current amplification factor of the transistor is

  (a) 100
  (b) 200
  (c) 300
  (d) 400

(34) Refer to Fig.Q13. The purpose of C_1 is to provide

  (a) r.f. filtering
  (b) interference suppression
  (c) inter-stage coupling
  (d) bias decoupling

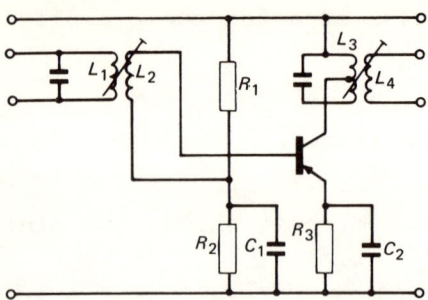

Fig.Q13

(35) Refer to Fig.Q13. The purpose of the tapping point on L_3 is to

    (a) increase the bandwidth
    (b) improve the selectivity
    (c) tune the amplifier
    (d) provide feedback

(36) Refer to Fig.Q14. The counter shown will divide by

    (a) 6
    (b) 8
    (c) 14
    (d) 16

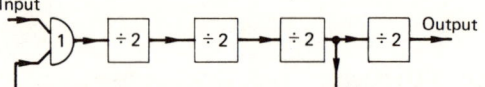

Fig.Q14

(37) An n-channel fet is working with class C bias with a sinusoidal input voltage. During which part of the cycle will drain current flow?

    (a) The peak of the positive half cycle
    (b) The whole of the positive half cycle
    (c) The whole of the positive and negative half cycle
    (d) The peak of the negative half cycle

(38) In the operational amplifier shown in Fig.Q15 the resistor R has a value of

    (a) 1 kΩ
    (b) 10 kΩ
    (c) 100 kΩ
    (d) 1 MΩ

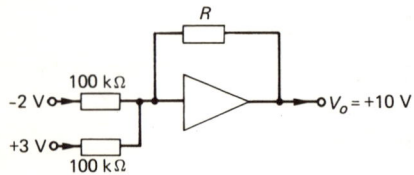

Fig.Q15

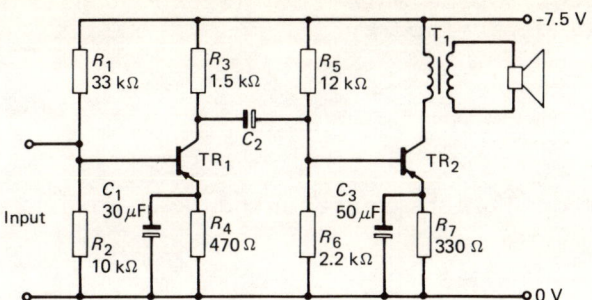

Fig.Q16

(39) Refer to Fig.Q16.   The purpose of C_3 is

    (a) coupling
    (b) decoupling
    (c) i.f. smoothing
    (d) d.c. biasing

(40) Refer to Fig.Q16.   A suitable capacitor C_2 is

    (a) 20 pF
    (b) 0.01 µF
    (c) 0.1 µF
    (d) 10 µF

(41) Refer to Fig.Q16.   If the collector of TR_1 has a
voltage of 4.5 V, TR_1 current is

    (a) 5 mA
    (b) 2 mA
    (c) 1 mA
    (d) 0.5 mA

(42) Refer to Fig.Q17.   The steady stage of the monostable
is

	V_1	V_2
(a)	10 V	10 V
(b)	0 V	10 V
(c)	10 V	0 V
(d)	0 V	0 V

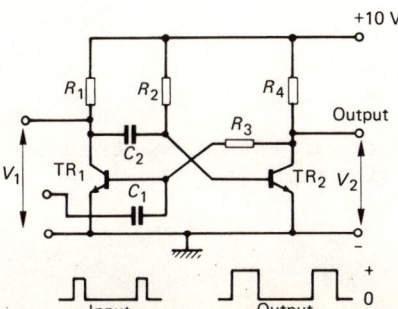

Fig.Q17

(43) Refer to Fig.Q17. The mark time (width of pulse) of the output is determined by

(a) R_1-C_2
(b) R_2-C_2
(c) C_1-R_3
(d) R_3-R_4

(44) A power loss of 3 dB is represented by a power ratio of

(a) $\frac{1}{3}$
(b) $\frac{1}{2}$
(c) 2
(d) 3

(45) Pentodes are preferred for high frequency amplifiers because

(a) pentodes use the Miller effect
(b) low-resistance anode loads can be used
(c) interelectrode capacitance effort is reduced
(d) low value of coupling capacitor

(46) A 1 kW domestic heater is used for 2 hours every day. If the cost of electric energy is 3p per unit the weekly cost is

(a) 3p
(b) 6p
(c) 21p
(d) 42p

(47) A fet amplifier in which drain current flows for less than half a cycle of the input signal waveform is termed to be operating in class

(a) A
(b) B
(c) AB
(d) C

(48) Refer to Fig.Q18. The characteristic curve shown is that of a

(a) thyratron
(b) unijunction transistor
(c) SCR
(d) tunnel diode

(49) Refer to Fig.Q18. The part of the curve that represents a negative resistance is

(a) OP
(b) OA
(c) AB
(d) BC

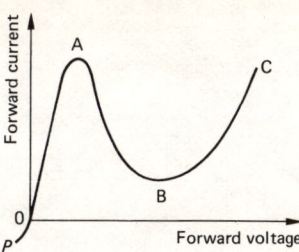

Fig.Q18

(50) A receiving aerial should be matched to a transmission
cable in order to

(a) increase the impedance of the aerial
(b) form standing waves
(c) draw minimum power from the aerial
(d) transfer maximum power to the aerial

(51) One of the reasons why ferrite rods are used for
radio receiver aerials is that they have

(a) low resistivity
(b) high permeability
(c) low Q factor
(d) high eddy current loss

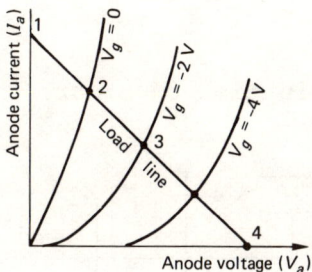

Fig.Q19

(52) Refer to Fig.Q19. The characteristics shown are
known as

(a) input for a triode
(b) output for a triode
(c) input for a pentode
(d) output for a pentode

(53) Refer to Fig.Q19. To obtain class B operation the
operating point should be at

(a) 1
(b) 2
(c) 3
(d) 4

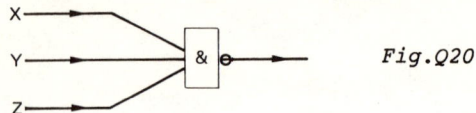

Fig.Q20

(54) Refer to Fig.Q20. If the inputs X, Y and Z to the gate are as shown, the output will be

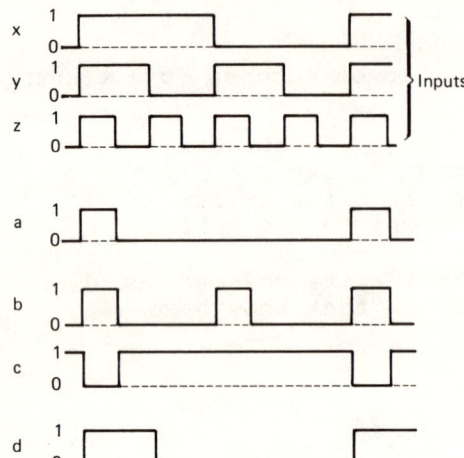

(55) Refer to Fig.Q21. Which combination of binary inputs would produce the output shown?

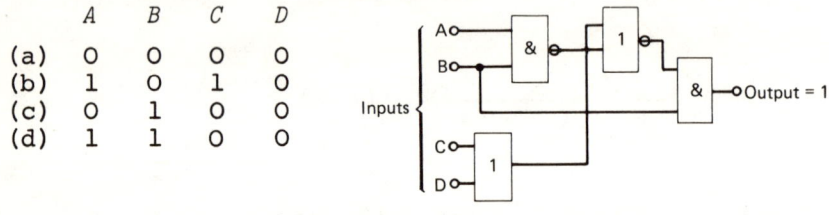

	A	B	C	D
(a)	0	0	0	0
(b)	1	0	1	0
(c)	0	1	0	0
(d)	1	1	0	0

Fig.Q21

(56) Refer to Fig.Q22. If the driven drum B has an arc of movement of 180° how many turns of the drive spindle A are required to rotate B from minimum to maximum?

(a) 5
(b) 10
(c) 25
(d) 50

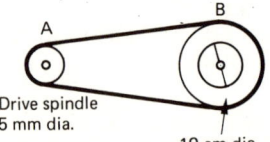

Drive spindle
5 mm dia.

10 cm dia.

Fig.Q22

(57) A drum rotates four complete revolutions in one second. The angular velocity in radians per second is

  (a) 4
  (b) 8
  (c) 4π
  (d) 8π

(58) Which one of the following combinations gives a resultant resistance of 75Ω

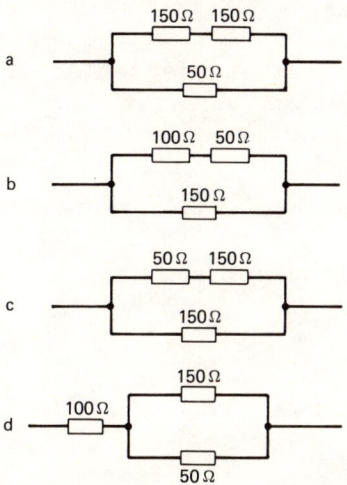

(59) Refer to Fig.Q23. The average voltage of the sine wave shown is

  (a) 0 V
  (b) - 5 V
  (c) 5 V
  (d) 10 V

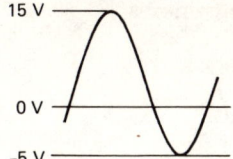

Fig.Q23

(60) The reactance X_c of a capacitor C is given as

  (a) ωC
  (b) $1/\omega C$
  (c) ω/C
  (d) C/ω

(61) Applying negative feedback to an amplifier will

  (a) lower the gain
  (b) increase the gain
  (c) produce oscillations
  (d) reduce the bandwidth

(62) Forward a.g.c. uses transistors whose gain

    (a) increases with current
    (b) decreases with current
    (c) stays constant irrespective of the current
    (d) is very low

(63) Directional aerials use a reflector which is

    (a) shorter than the dipole
    (b) longer than dipole
    (c) shorter than the director
    (d) same length as the dipole

(64) A 20 000 Ω/V voltmeter has a full-scale deflection current of

    (a) 20 μA
    (b) 50 μA
    (c) 20 mA
    (d) 50 mA

(65) Refer to Fig.Q24. The no load voltage is

    (a) 240 V
    (b) 350 V
    (c) 480 V
    (d) 700 V

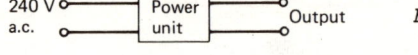

Fig.Q24

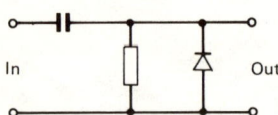

Fig.Q25

(66) Refer to Fig.Q25. The output waveform produced by the circuit shown is

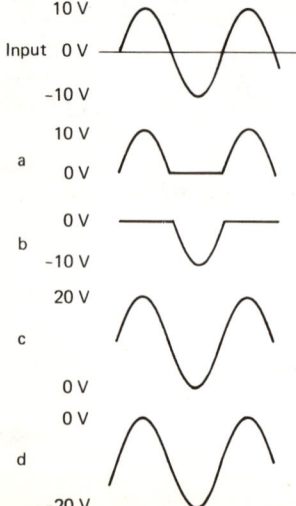

(67) Interstage coupling capacitors in a.f. circuits employing unipolar transistors have much smaller values than those using bipolar transistors because bipolar transistors have

(a) low input impedance
(b) high input impedance
(c) high leakage current
(d) low gain

(68) Refer to Fig.Q26. The maximum safe voltage that may be applied across A and B is

(a) 5 V
(b) 10 V
(c) 15 V
(d) 20 V

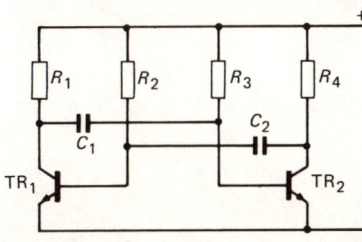

C_1 C_2

A o———||————————||———o B   *Fig.Q26*

0.1 μF   0.2 μF
(WV 10 V)   (WV 20 V)

(69) Refer to Fig.Q27. The name given to the oscillator circuit shown is

(a) Colpitts
(b) Hartley
(c) astable multivibrator
(d) bistable multivibrator

R_1 R_2 R_3 R_4

C_1 C_2

TR₁   TR₂

Fig.Q27

(70) Refer to Fig.Q27. If R_2 is half the value of R_3 and C_1 is half the value of C_2 the mark-space ratio of the current will be

(a) 1 : 2
(b) 1 : 1
(c) 2 : 1
(d) 4 : 1

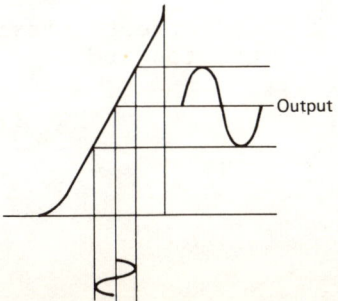

Output

Fig.Q28

Input

(71) Refer to Fig.Q28. This shows a transfer characteristic with input and output signals. The amplifier
is operating in class

(a) A
(b) B
(c) AB
(d) C

(72) The torque produced by a force of 2 N acting at a
radius of 100 cm is

(a) 0.1 Nm
(b) 0.2 Nm
(c) 0.5 Nm
(d) 2 Nm

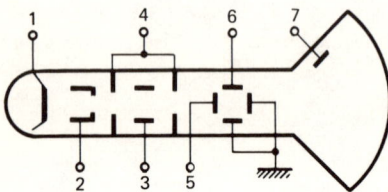

Fig.Q29

(73) Refer to Fig.Q29. The brightness and focus electrodes are

(a) 1,2
(b) 2,3
(c) 2,6
(d) 1,7

(74) A device which emits electrons when light falls upon
it is known as a

(a) photoconductive cell
(b) photovoltaic cell
(c) photo diode
(d) photo-emissive cell

(75) The impedances presented by a field effect transistor
used in the common source configuration are

(a) low input and low output
(b) low input and high output
(c) low input and medium output
(d) high input and medium output

A5 Practical Applications Question Paper

(1) Complete the table of SI units.

QUANTITY	UNIT NAME
Energy	
	watt
	tesla
impedance	
electromagnetic force	

(2) Name the devices represented by the symbols shown in Fig.Q30.

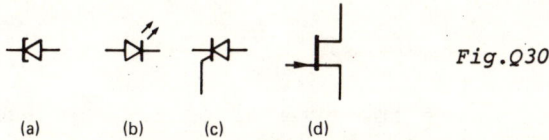

Fig.Q30

(a)   (b)   (c)   (d)

(3) Sketch the current/voltage characteristics curve for silicon and germanium diodes.

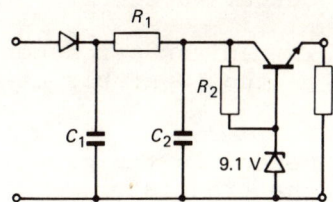

Fig.Q31

(4) Refer to Fig.Q31. What would be the effect on the output if C_2 went open circuit?

(5) Refer to Fig.Q31. In what configuration is the transistor connected?

(6) Refer to Fig.Q31. Assuming the transistor is silicon, what is the output voltage?

(7) Refer to Fig.Q31. What would the output be if the zener goes short circuit?

(8) If the output of an amplifier is 1 W when the input is 1 mW, state the gain in dB.

(9) Why are class C amplifiers used in oscillators?

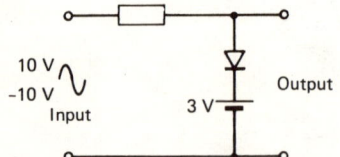

10 V

−10 V

Input

3 V

Output

Fig.Q32

(10) Refer to Fig.Q32. Sketch the output voltage waveform when the input is the sinewave shown.

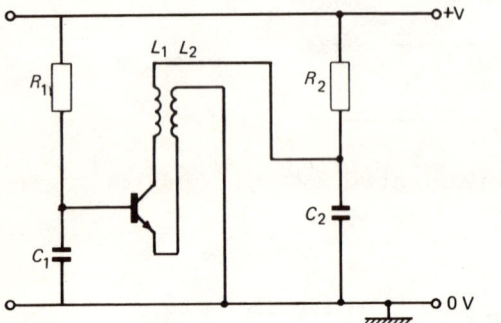

Fig.Q33

(11) Refer to Fig.Q33. Name the type of oscillator shown.

(12) Refer to Fig.Q33. Which components determine the frequency of the output?

(13) Refer to Fig.Q33. Sketch the output waveform.

(14) What is the angular velocity, in radians per second, of a record which is rotating on the turntable at 45 rev/min?

(15) Show how a diode should be connected in the circuit (Fig.Q34) in order to protect the transistor.

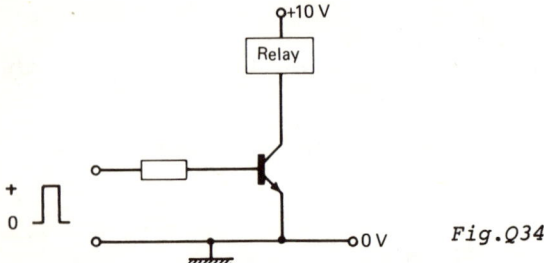

+10 V

Relay

+

0

0 V

Fig.Q34

(16) Refer to Fig.Q34. If the relay coil has a resistance of 120 Ω what power will be dissipated in the transistor when the collector current is 50 mA?

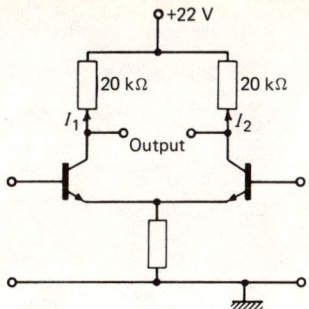

Fig.Q35

(17) Refer to Fig.Q35. If I_1 = 500 μA and I_2 = 450 μA, calculate the output voltage.

(18) With no input applied to the a.f. amplifier shown in Fig.Q36 state, giving reasons, the effect on the drain current if

    (a) R_2 becomes open circuited
    (b) R_1 becomes open circuited.

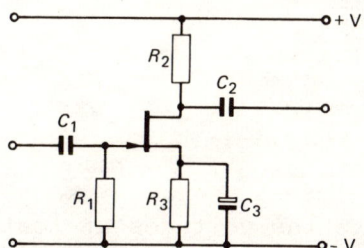

Fig.Q36

(19) Refer to Fig.Q36. What is the effect of C_3 going open circuit on

    (a) d.c. (quiescent) condition
    (b) the output signal.

(20) Refer to Fig.Q36. Give typical values for

    (a) C_1
    (b) C_3

(21) The curve shown in Fig.Q37 represents the regulation of a power supply having a full load current of 25 mA. Calculate the percentage regulation.

(22) Refer to Fig.Q37. State the no load voltage.

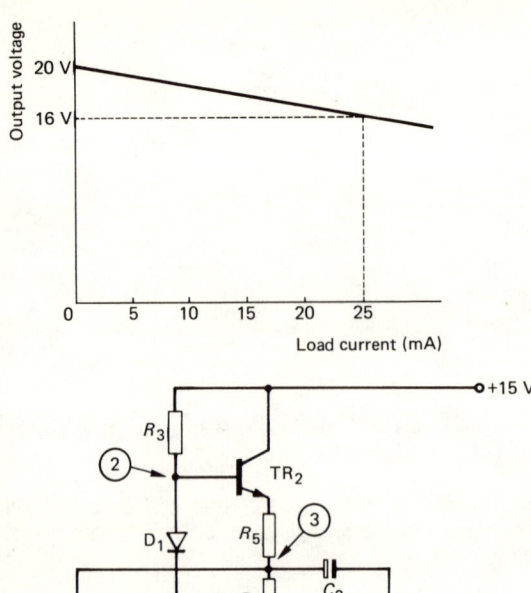

Fig.Q37

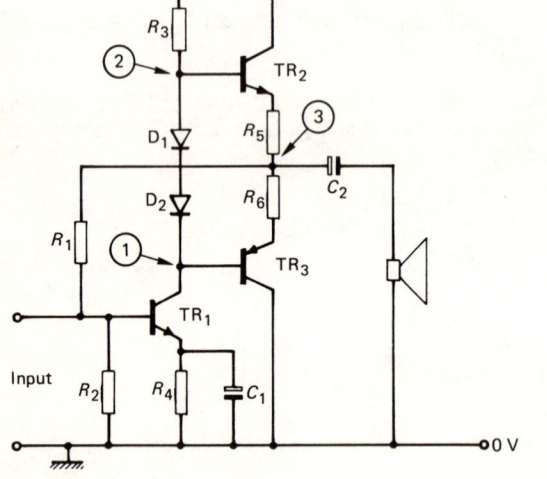

Fig.Q38

(23) Refer to Fig.Q38. Estimate the voltages at test points 1, 2 and 3.

(24) Refer to Fig.Q38. State the purpose of D_1 and D_2.

(25) Refer to Fig.Q38. State typical values for R_5, R_6 and C_2.

(26) State the reason why electromagnetic deflection is *not* used in cathode ray oscilloscopes.

(27) State the reason for giving field effect transistors the name "unipolar" transistors.

(28) State the purpose of a matching transformer used in a.f. amplifiers.

(29) State the name given to the impedance of an infinitely long transmission line.

(30) Give two reasons why negative feedback is used in amplifiers.

Answers (Appendix 4)

(1)	*b*	(26)	*d*	(51)	*b*
(2)	*c*	(27)	*a*	(52)	*b*
(3)	*c*	(28)	*c*	(53)	*d*
(4)	*c*	(29)	*c*	(54)	*c*
(5)	*b*	(30)	*a*	(55)	*d*
(6)	*b*	(31)	*d*	(56)	*b*
(7)	*c*	(32)	*d*	(57)	*d*
(8)	*d*	(33)	*a*	(58)	*b*
(9)	*c*	(34)	*d*	(59)	*c*
(10)	*b*	(35)	*b*	(60)	*b*
(11)	*b*	(36)	*c*	(61)	*a*
(12)	*d*	(37)	*a*	(62)	*b*
(13)	*c*	(38)	*d*	(63)	*b*
(14)	*d*	(39)	*b*	(64)	*b*
(15)	*a*	(40)	*d*	(65)	*b*
(16)	*b*	(41)	*b*	(66)	*d*
(17)	*b*	(42)	*c*	(67)	*a*
(18)	*d*	(43)	*b*	(68)	*c*
(19)	*c*	(44)	*b*	(69)	*c*
(20)	*c*	(45)	*c*	(70)	*b*
(21)	*c*	(46)	*d*	(71)	*a*
(22)	*d*	(47)	*d*	(72)	*b*
(23)	*b*	(48)	*d*	(73)	*b*
(24)	*c*	(49)	*c*	(74)	*d*
(25)	*b*	(50)	*c*	(75)	*d*

Answers (Appendix 5)

(2) (*a*) Tunnel diode

    (*b*) Light emissive diode

    (*c*) Silicon controlled rectifier

    (*d*) n-channel fet

(4) Increased ripple

(5) Emitter follower

(6) 8.5 V

(7) 0 V

(8) 30 dB

(9) High efficiency

(11) Blocking

(12) R_1-C_1

(14) 1.5π rad/s

(16) 200 mW

(17) 1 V

(18) (*a*) Zero drain current

    (*b*) Increases due to the absence of reverse bias

(19) (*a*) No effect

    (*b*) Low

(20) (*a*) 0.1 µF

    (*b*) 10 µF

(21) 20%

(22) 20 V

(23) Test point 1: 7.6 V

    Test point 2: 7.4 V

    Test point 3: 7.5 V

(24) To overcome crossover distortion

(25) 2.2 Ω, 2.2 Ω, 300 µF

(26) Low frequency response

(27) FETs use one type of carrier

(28) Maximum power transfer

(29) Characteristic impedance

(30) Stability and increased bandwidth

INDEX

NOTES

NOTES

NOTES